Origin of Life and the Universe

The Science of Creation

Paul Garnett

Origin of Life and the Universe
The Science of Creation
By Paul D. Garnett
M.A., O.D.,D.E.S.(Oxon)

Paul Garnett

Paul Garnett

A cknowledgements

GRAPHICS: Anna Garnett

GRAPHICS CONTRIBUTOR
Donald McCain
Jayne Bauer

COMPUTER TECHNICAL ADVISOR
Steve Small
Christine Diepenbrock

TECHNICAL ADVISOR: S.E.Gulino, PhD.

Origin of Life and the Universe
The Science of Creation

By Paul D. Garnett

In this book you will find . . .

DAY 1 (Genesis I)

Darkness   EREV   BOKER

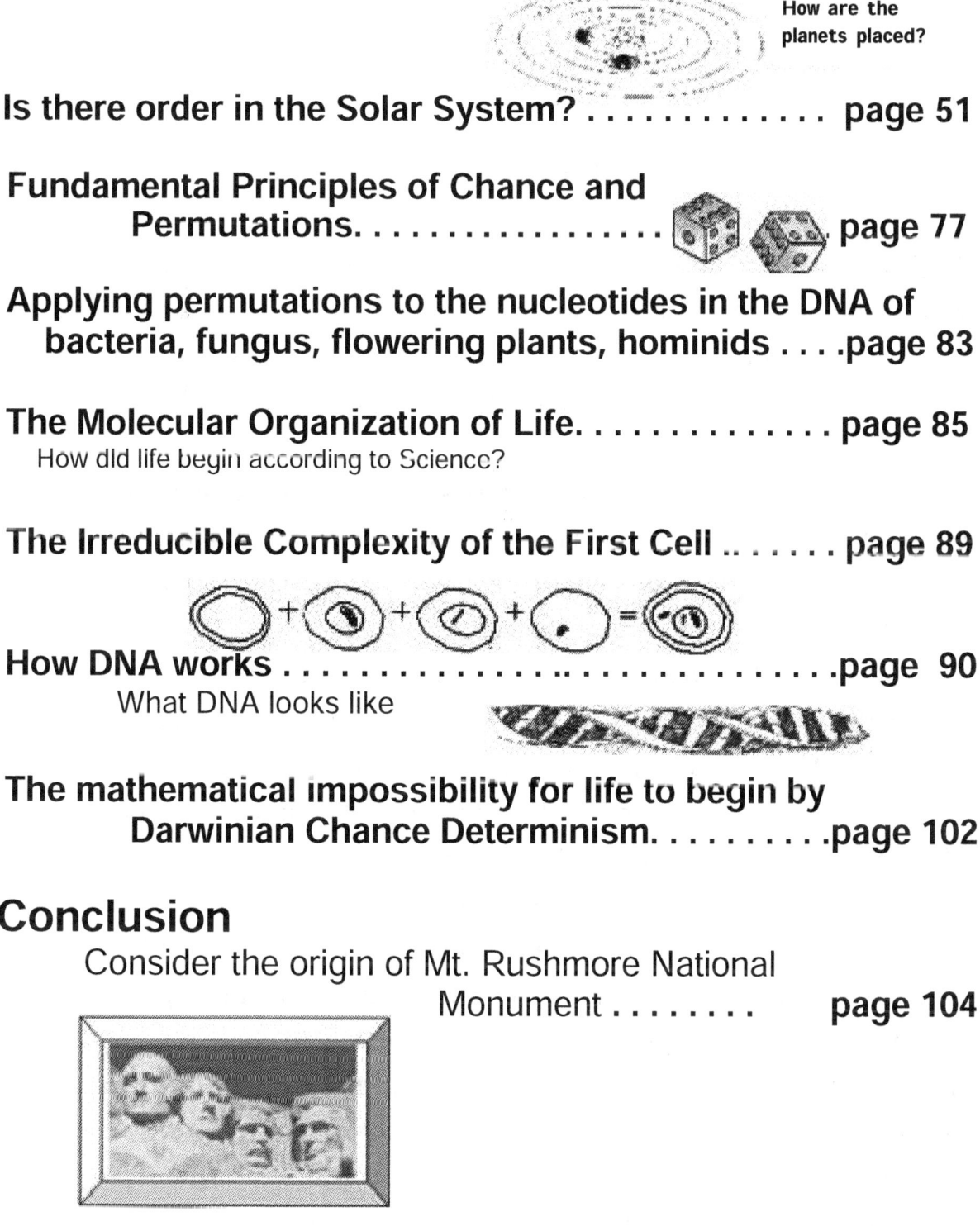

How are the planets placed?

Conclusion

Genesis and the Cosmic Clock

Genesis means *original source* in Greek. It is derived from the Greek verb *gen-nao*, which means *birth*, or *to give birth to*, and comes from the English translation of the first five books of the Bible. It was written in ancient Hebrew and then translated into Greek. Why Greek? The Hebrew historian Josephus tells us that sometime between 285 and 247 BC the Pharaoh Ptolemy Philadephus ordered the Greek translation of these ancient Hebrew writings.

The task was assigned to 72 priests, six from each of the twelve tribes of Israel. It was to be completed in 72 days; hence the name Septuagint, which means 72. From this Greek translation, and then Latin, and finally to English, we get our present day Bible. No one can read this amazing book without being astonished at its coherence, although there were more than 40 different writers spanning a period of 1600 years. The writers wrote in prisons, palaces, deserts, pastures, cities, mountains, lakes, forests and temples, and were separated by hundreds of miles and hundreds of years. They came from all walks of life. There were princes, poets, fishermen, herdsmen, doctors, lawyers, musicians, herdsmen, farmers, kings, and generals. One would expect the result to be a mixed bag of ideas and concepts; but instead, there is an amazing unity and continuity as the book tells about God's dealings with mankind. The opening book of the Bible tells us about the origin of the Universe, and the Earth, and how life began. This text was written thousands of years ago, before the advent of the modern discipline of science.

Has science replaced the Bible as the authority we should teach our children about our origins? This book addresses the apparent conflicts in Bible dogma and modern scientific theories.

It is not an accident that creation of light was the first step in creating the Universe[1], since light (radiation) can step out of the realm of energy and change into matter. When that happens, it enters into the measurable world of time and space following Einstein's famous equation of $E=mc^2$.

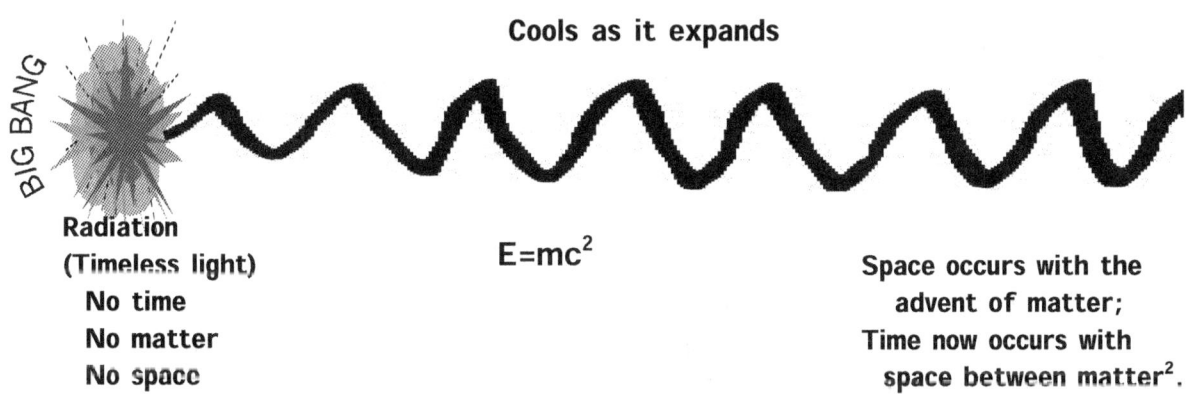

Cools as it expands

BIG BANG

Radiation
(Timeless light)
No time
No matter
No space

$E=mc^2$

Space occurs with the
advent of matter;
Time now occurs with
space between matter[2].

To understand the human flow of time, we must look into origin of light, thus the origin of matter, and from these the expansion of space.[1] If we are to get an accurate measure of time, what shall we use for a clock? We must select a clock that uses the same standard throughout the Universe that includes light rays: electromagnetic radiation. A common mistake in studying the Biblical ages in Genesis is to view the universe from a specific location, rather than selecting a reference point that includes the entire Universe.

To find our cosmic clock, we must search for a clock that measures time, not between LOCATIONS but between EVENTS that occurred after the initial explosion of that tiny grit of primordial material into the vast expanses of galaxies and gases that we see in space today. We cannot use the light that originates from distant stars and galaxies because these come from DIFFERENT LOCATIONS, and have different time frames. [2]

---

[1] "Space is the physical entity in which matter exists. Matter is controlled by physical laws which brings us to the problem of cause and effect. Effects are separated from cause by time, if there were no time before the Universe came into existence, then there was no space for the effects of space to take place. Matter must precede space. . . . (J.Maddox: "*Down with the Big Bang*", Nature, **340:425**, 1989. (John Maddox is Editor of *Nature*.)

[2] " . . .Each planet, each star, each location within our universe has its own unique gravitational potential, its own relative velocity, and therefore, its own unique rate at which the local proper time passes. . ." (Ref.

However, there is another radiation that has been present from the creation of the Universe. This is the "echo" of the Big Bang, and it is called the Cosmic Background Radiation (CBR). It fills the entire Universe WITHOUT BEING RELATED TO ANY PARTICULAR LOCATION. (Ref.45)

This radiation is the heat left over after the explosion of the Big Bang, and it tells us how long it has taken for the Universe to "cool" down to its present temperature: 2.78 degrees K, or -270 degrees C, or -450 degrees F. This cooling "stretches", or expands the wavelengths, and thus its frequency.

Cosmic Clock

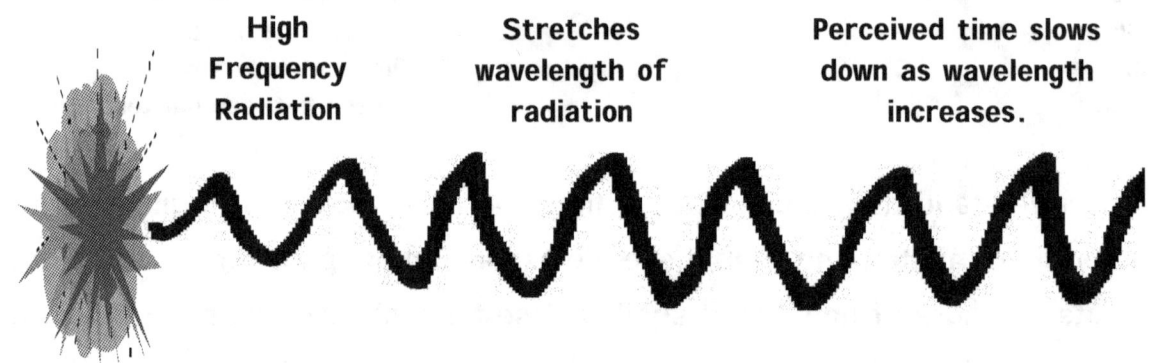

| High Frequency Radiation | Stretches wavelength of radiation | Perceived time slows down as wavelength increases. |

Contains
Intense heat
(energy)

Cools as it expands

Slows conventional
(biological) time

BETWEEN SPECIFIC LOCATIONS

In addition to the "Stretching" Radiation, there is another important effect of the expansion of the Big Bang, and that is the formation of SPACE that takes place by the conversion of energy into particles of matter following Einstein's equation of $e=mc^2$. (Ref.14) With the appearance of matter our Cosmic Clock starts. (Ref.29)

Hebrew scholars (Nahmanides) points this out to us by showing the first word in the Bible is BERAESHEET, which means: "In the beginning of . . .". Notice the extra English word, "of". In the beginning of . . . WHAT?

P.J.E. Peebles, *Principles of Physical Cosmology*, Princeton University Press, Princeton, 1993 pp 71, 91, 96, 135).

The answer is TIME, God's cosmic time. We know that because all electromagnetic radiation, (such as the light rays that permit us to read these pages) exists in a state of everlasting NOW, when there is no passage of time.[3] Thus, it is God's time. We mortals have some difficulty in grasping this, since everything we know is related to a world we can see, feel, and touch. Thus, we should be able to grasp the concept that time ("In the beginning of") begins when matter comes into being.

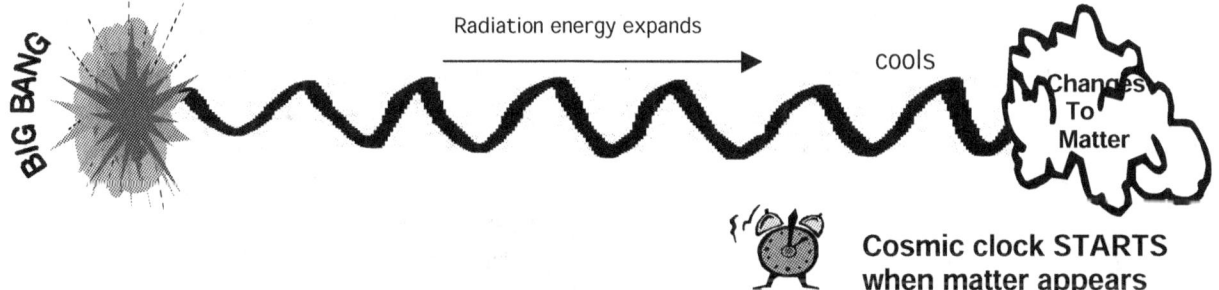

BIG BANG

Radiation energy expands

cools

Changes To Matter

Cosmic clock STARTS when matter appears

Continuing on with the Hebrew text in Genesis 1:2: (Ref.30) ". . . and the earth was "TOHU" and "BOLU". The words have usually been translated: ". . . and the earth was formless and void." But both the Talmud and Nahmanides tell us that BOLU also means: "filled with building blocks of matter". (Ref.30)   At about what time, according to science, was this condition present as measured by Earth/Sun years? By our calculation using the expression for the natural log e and the ratio of the temperature CBR^6 (NOW: 2.73K) to the calculated temperature of the Big Bang (THEN: 10.9 x 10^{12} K), . . . scientists estimate the Earth to be about 9 billion Earth years old.

Since time begins with the appearance of matter, our Cosmic Clock begins with BOLU ("Building Blocks of Matter): the instant that matter was formed from the cooling radiation of the Big Bang as it expanded outward. In the Physics lab we can calculate the temperature, and hence the frequency of the radiation at the moment time began in the Universe.   This has been defined as "Quark Confinement" or that instant Quarks were confined by nuclei to form atoms (See footnote #6).

---

[4] ". . . Radiant energy, such as light rays, which permit vision, exist in a state that might be described as an "eternal now. . . in a state in which time does not pass." (Ref: G. Schroeder. *The Science of God,* The Free Press, N.Y., 1997, pp.56.)

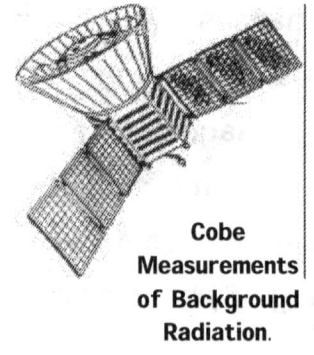

Cobe Measurements of Background Radiation.

One of the most extensive space experiments ever attempted by NASA was to investigate a mysterious radiation from space. It was proposed perhaps the mysterious radiation was the echo of the Big Bang. It took NASA 12 years to plan and carry out a space experiment to discover what the radiation was, but it took only eight minutes to give the results after the space satellite began to transmit data back to Earth[4]. The data exactly conformed to the predicted properties of the cosmic background radiation for the Big Bang. (Ref.18)

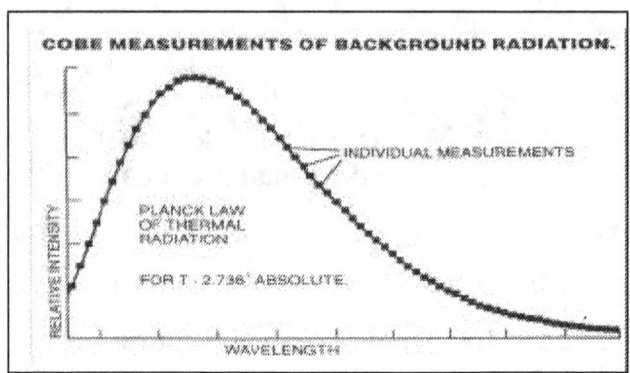

These data show that the cosmic background radiation from the Creation Event fits the spectral profile of nearly a perfect radiator to a precision of 0.03%. This was proof that the satellite detectors were looking at the initial, infinitely hot, dense state of a newly created universe.

Because the present wavelengths "stretched" as they cooled, we can compare the Cobe measurements NOW with the calculated wavelengths THEN. This is about a million million times hotter than the present temperature of outer space (2.87 degrees K)[5]. Our present radiation has been stretched a million, million times greater from what it was originally. Time thus has slowed down a million million times. In physics, this is known as time dilation (Ref. 19)[6], the basis for an accurate cosmic clock using CBR (Cosmic Background Radiation).

---

[5] "Launched in 1989, the instruments took only 8 minutes to verify the conclusions based on the 1964 measurements of Penzias and Wilson." (Ref: J.P. McEvoy & O. Zarate, Stephen Hawking for Beginners, Icon Books Ltd., U.K. p.168)

[6] The radiation from quark confinement has been stretched a million millionfold. Its red shift, z, as observed today is 10. That stretching of the light waves has slowed the frequency of the cosmic clock . . . by a million million. (Ref: G. Schroeder. *The Science of God*, The Free Press, N.Y., 1997 pp. 57)

[6] This million million also applies to proper rates of events as one sees by the application of a sequence of Lorentz time-dilations factors. (Ref: Peebles, op.cit, p.96)

Linear Time on Earth

.Science looks <u>BACKWARD</u>
From now to the Big Bang using Cobe calibrated radiation as a standard of measurement.

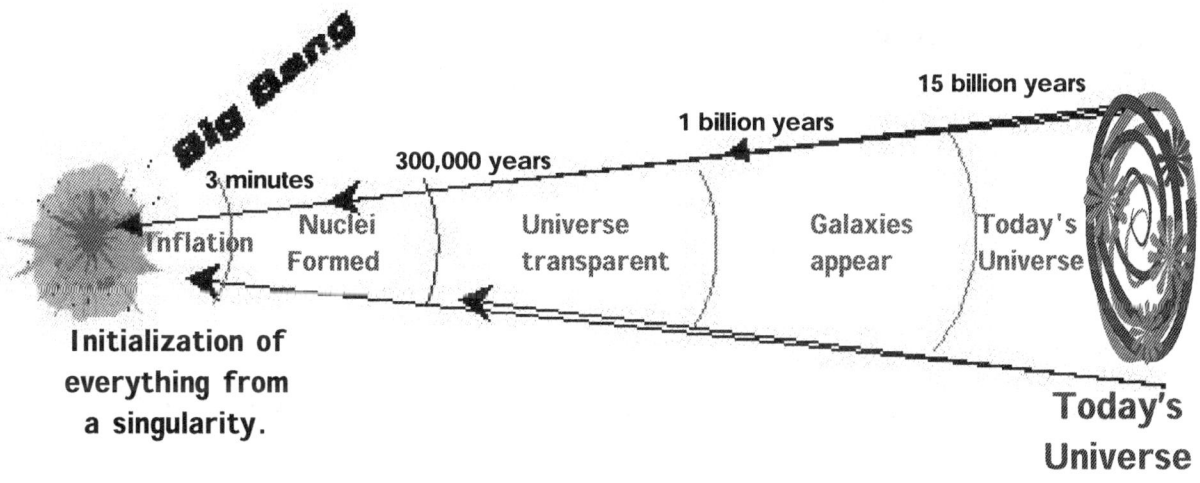

15 billion years
1 billion years
300,000 years
3 minutes
Inflation
Nuclei Formed
Universe transparent
Galaxies appear
Today's Universe

Initialization of everything from a singularity.

Today's Universe

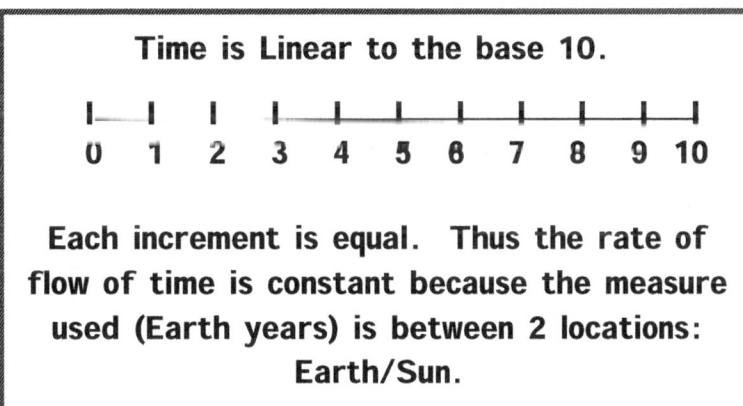

Time is Linear to the base 10.

0 1 2 3 4 5 6 7 8 9 10

Each increment is equal. Thus the rate of flow of time is constant because the measure used (Earth years) is between 2 locations: Earth/Sun.

Exponential Time of Universe as shown in the Bible

Rate of Change in the Universe is UNEQUAL
because God's measure of Time is between
events not *locations.*

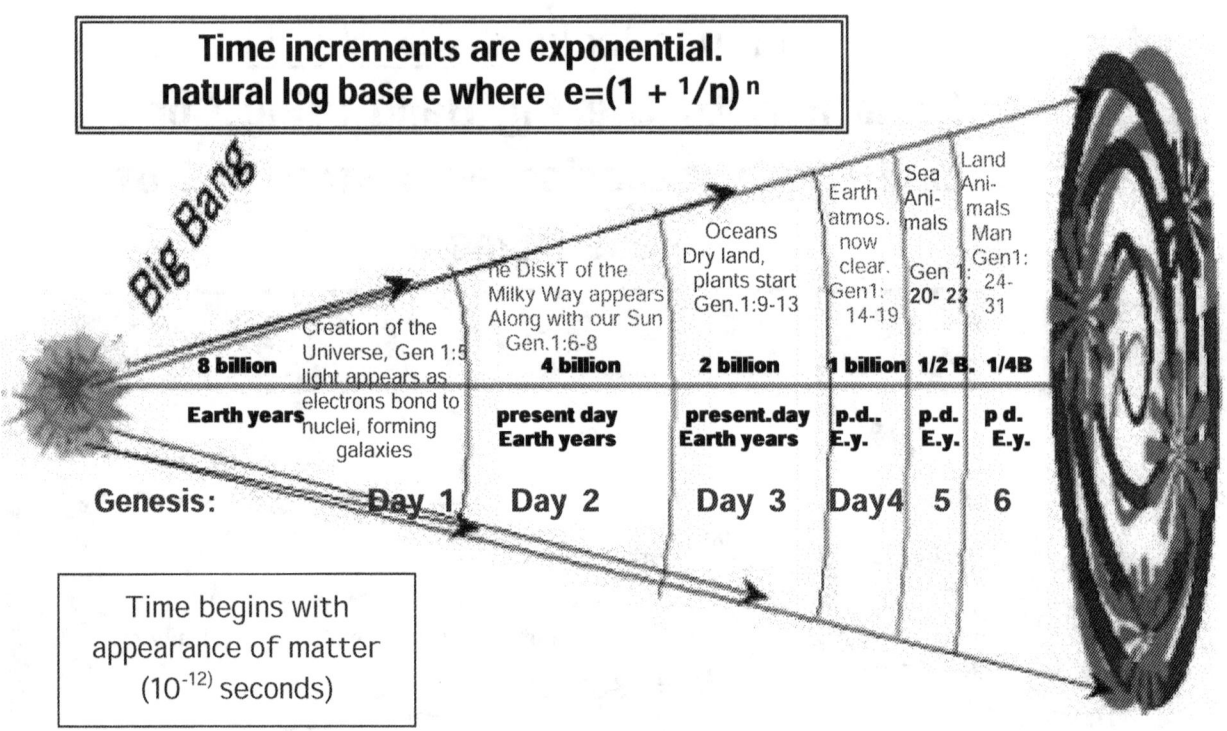

Time increments are exponential.
natural log base e where $e=(1 + {}^1\!/n)^n$

Big Bang

| | | Earth atmos. now clear. Gen1: 14-19 | Sea Ani-mals Gen 1: 20- 23 | Land Ani-mals Man Gen1: 24-31 |

Oceans Dry land, plants start Gen.1:9-13

he DiskT of the Milky Way appears Along with our Sun Gen.1:6-8

Creation of the Universe, Gen 1:5 light appears as electrons bond to nuclei, forming galaxies

| 8 billion | 4 billion | 2 billion | 1 billion | 1/2 B. | 1/4B |
| Earth years | present day Earth years | present.day Earth years | p.d.. E.y. | p.d. E.y. | p d. E.y. |

Genesis: Day 1 Day 2 Day 3 Day4 5 6

Time begins with
appearance of matter
(10^{-12}) seconds)

The Bible account looks FORWARD from the beginning of time to the creation of Adam, not of hominids.

Note that, unlike science, the rate of change is exponential, not linear. The rate is an expansion of the Natural Log e, which describes a form found throughout Nature more frequently than any other form. Here on Earth this mathematical expansion describes among many other things, the graceful curve of the nautilus shell, the distribution of seeds in a sunflower, the curve of animal tusks, and the shape of a spider's web. In outer space it describes the spread of stars in distant spiral galaxies. On Wall Street, it is the formula for compound interest, mathematically stated: $e=(1+{}^1\!/n)^n$. This expression is known as the Golden Section, or Fibonacci series, and is also used as a discrimination between inorganic and organic systems. (Ref.15.1)

For big numbers, n = 2.71827 . . .   The sequence goes on forever, but never repeats nor becomes greater than 2.71828. It is one of God's favorite numbers, which Genesis uses to measure time increments for the *Six Days of Creation*.

To study this we will be comparing:

Rate of <u>Linear Time (Base log 10) since Creation (About 16 billion Earth</u> years)
<u>Exponential Time (natural log base e) of Biblical 6 Days (Earth years)</u>

$\equiv$ (equals)

Temperature at the moment of the Big Bang (10.9 x 10^{12} degrees K)THEN
Temperature of the Background Radiation NOW (2.73 degrees K)NOW

OR

One million million   (1,000,000)(1,000,000)(temperature of Big Bang)THEN
1 (temperature of CBR) NOW

The ratio between Earth Time (24 hour days) and Cosmic Time is about a million millions to one. This is based on the million millionfold stretching of light (radiation) as the Universe expands from the Big Bang to the present day.[7]

The number of present day Earth-years in each Genesis day is calculated by the ratio of comparison of the cosmic background radiation frequency NOW to cosmic background temperature at the time of the Big Bang [10.9 x 10^{12} degrees Kelvin (THEN) to 2.73 degrees Kelvin (NOW)] (Ref.19) and putting that ratio into the equation for the expansion of a spiral. (The Golden Section or Fabonnacci Series) For mathematics students, this involves some integral calculus which is explained in careful detail in Gerald Schroeder's text: *The Science of God* (pp.65-68), Simon & Schuster 1997.

---

[7] "Positive difference in velocity, gravity and the expansion of space . . ., all increase the wavelength of radiation. Since frequency of radiation is lowered in direct proportion to the increase in wavelength, the increased wavelength slows the perceived passage of time. This phenomena relates directly to differences in the flow of conventional -- biological time -- between specific locations. In addition the expansion of space alters the perception of time's flow. (Ref.: P.J.E. Peebles, op.cit., pp91-96)

To test our results we can now compare our calculated sequence with what science proposes as the history of our Universe, specifically the time frame. The scientifically calculated age of our Universe equals about 15.75 billion years, which comes close to the roughly estimated age of the oldest stars. (Ref.4)

From the Bible's perspective looking forward in time from start of Day One	From Science's perspective looking backward in time from the present
* *	* * * * * * * * * * ** * * * * * *
Day one 24 hours	8 billion years
Day two 24 hours	4 billion years
Day three 24 hours	2 billion years
Day four 24 hours	1 billion years
Day five 24 hours	1/2 billion years
Day six 24 hours	1/4 billion years
Total: Six 24 hour days	15-3/4 billion years

Remember that if we were to place a clock in any one of billions of different locations in the Universe, it would slowly tick away 15 3/4 billion Earth years while it recorded only six 24 hour days at some other place. In our own solar system, an "Earth clock" would be slower on the Moon, faster on the Sun simply because of gravity, to say nothing of velocity and mass. To double-check these conclusions, we need an independent calculation to confirm our results. Einstein has conveniently given us the key information to do this.

Einstein's space-time equations have shown us that the passage of time is related to the speed we are traveling. The faster we travel, the more time slows down, the ultimate speed being the speed of light.

Having shown this relationship, Einstein then showed that as radiant energy (light) expands so does space, and thus so does time. What happens when we apply Einstein's Time/Space Expansion Factor to the age of our Universe, calculated from the Big Bang?

Best estimate of age of Universe = 15.78 billion Earth years

$$\frac{15.58 \times 10^{9} \text{ (time in Earth years of Universe)}}{9.8625 \times 10^{12} \text{ (Einstein's Time/Space Expansion Factor)}} = .01642$$

(.01642 years)(365.25 days) = 5.999 24-hour Earth-days (6 DAYS OF CREATION)

(365.25 is number of 24-hour days in one Earth-year. The result is not exactly 6 days since the age of our Universe is an estimate.)

OK, the overall age fits.

What about the day to day events as described in Genesis, as compared with the fossil and paleontological record?

Paul Garnett

The Creation Model

The Seven Days of Creation in the Bible

How is this order coherent with Science?

Day 1: Heaven and Earth, Light (Genesis 1:3-5)

Day 2: Sky (Genesis 1:3-5

Day 3: Dry Land, Plant Life (Genesis 1:6-8)

Day 4: Sun, Moon, Stars, Timetable (Genesis 1:14-19)

Day 5: Animal Life (Genesis 1-20)

Day 6: Mankind (Genesis:1:24-31

Day 7: God rests (Genesis 2:1-2)

Day 1 God spoke and the Universe came into being (Big Bang).

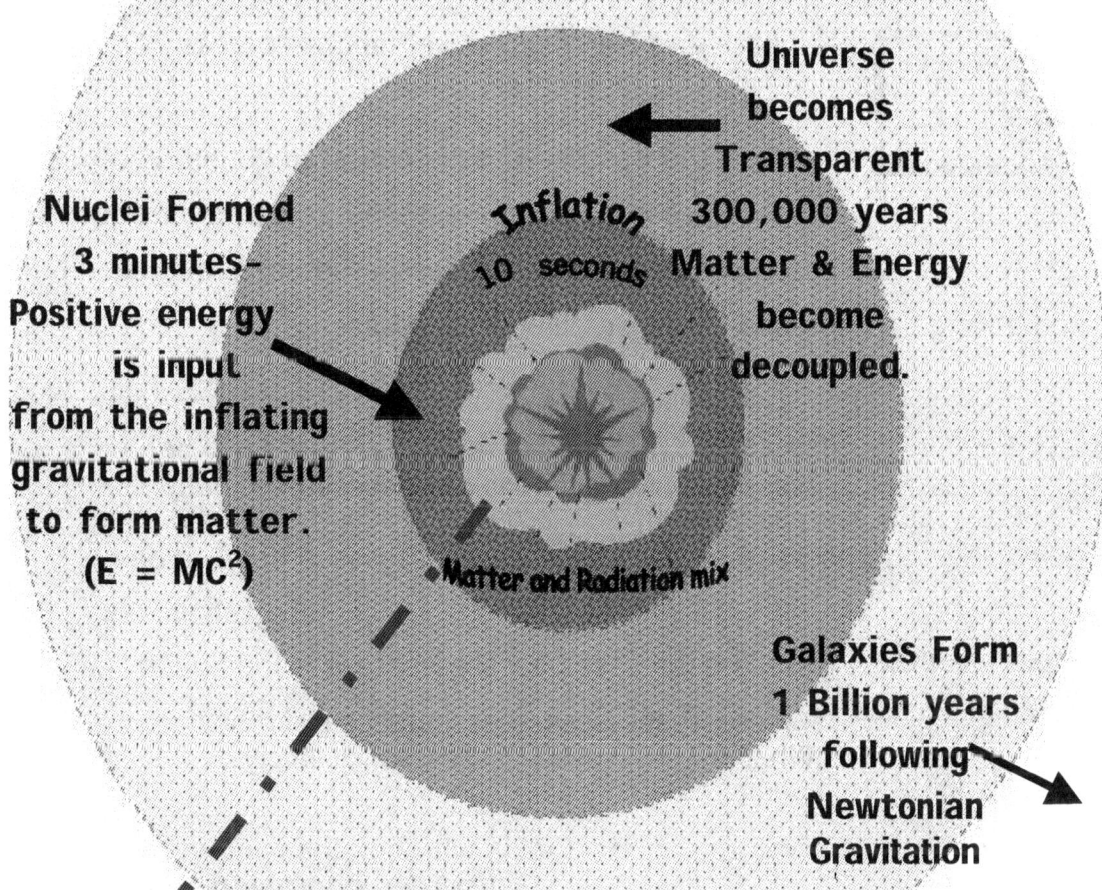

Universe becomes Transparent 300,000 years Matter & Energy become decoupled.

Nuclei Formed 3 minutes– Positive energy is input from the inflating gravitational field to form matter. $(E = MC^2)$

Inflation 10 seconds

Matter and Radiation mix

Galaxies Form 1 Billion years following Newtonian Gravitation

Background Radiation of Big Bang still persists.

INFLATION

Expansion from initial state smaller than proton to about 10M across in 1^{-12} seconds.

Four Fundamental Forces Occur in This Order

1. **Strong Nuclear (in place at moment of Big Bang)**
2. **Gravitational (10^{-43} seconds)**
3. **Weak Nuclear (10^{-10} seconds)**
4. Electromagnetic (10^{-35} seconds)

Today's Universe is 15.75+ Billion years

Creation of the Earth

Genesis 1:2

". . . and the earth was TOHU and BOLU."
The usual English translation: *". . . and the earth was formless and empty."*
Hebrew scholars (both the Talmud and Nahmanides) translate this passage as: *". . . and the earth was without form and was filled with building blocks of matter."* Ref 30)

Day 1

How did this chaotic collection of "building blocks and matter" form into our home, teeming with the fragile forms of life? The Bible does not give us a step by step recipe. Instead it tells us something that appears nowhere else in Scripture:

Genesis 1:2 *". . . . and the Spirit of the Eternal (I was, I am, I will be) was hovering over the darkness of the deep . . ."*

Genesis 1:3 *". . . and the evening and the morning were the first day . . "*
(English version)

How can this be if the Sun does not become visible until the Fourth Day?

To answer that question, we must consider first:
What do the original Hebrew words imply?

Hebrew word for *evening* ⟶ ***EREV*, which is the root word for disorder, confusion, mix-up. (Ref 30)**

Hebrew word for *morning* ⟶ ***BOKER*, which is the root word for orderly, organized, sequential (Ref.30)**

**   Please note that Genesis tells us:
Evening comes first *(disorder);*
Morning comes second *(orderly, organized)*

Our usual understanding of a day on Earth starts with morning, not evening. The more logical and appropriate translation would be something to the effect that each day started with disorder and concluded with order and organization, rather than telling us about evening and morning; however, Jews count the Sabbath from evening to morning.

Secondly, prior to all of this, " . . .*the Spirit of God hovers over the darkness of the deep.*"   Genesis 1:2

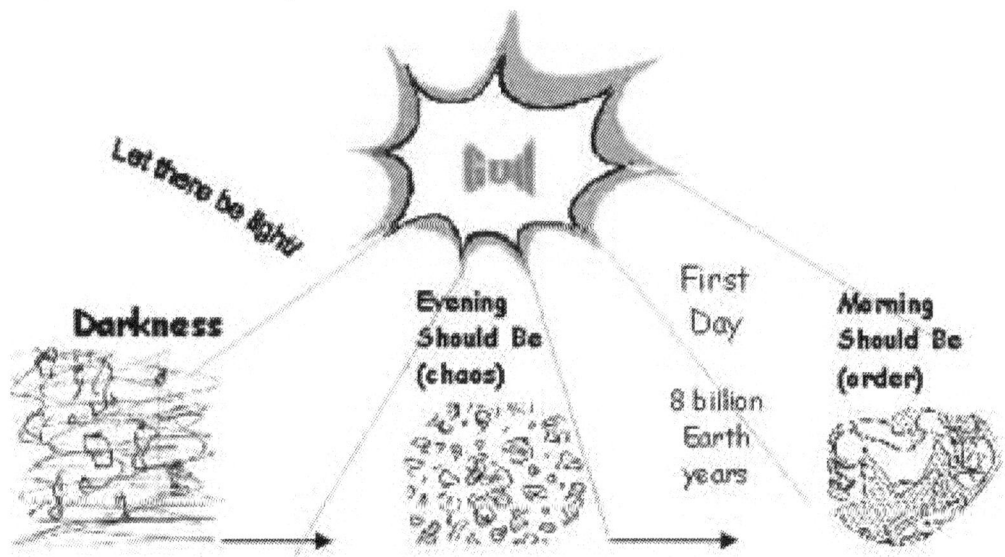

God put the "building blocks" of the Earth (and the Universe) together by instituting the precise Fundamental Forces (Laws) of the Universe . . . a divine sorting plan. These 4 Forces are God's rules for bringing order out of chaos, although Jews count the Sabbath from evening to morning.

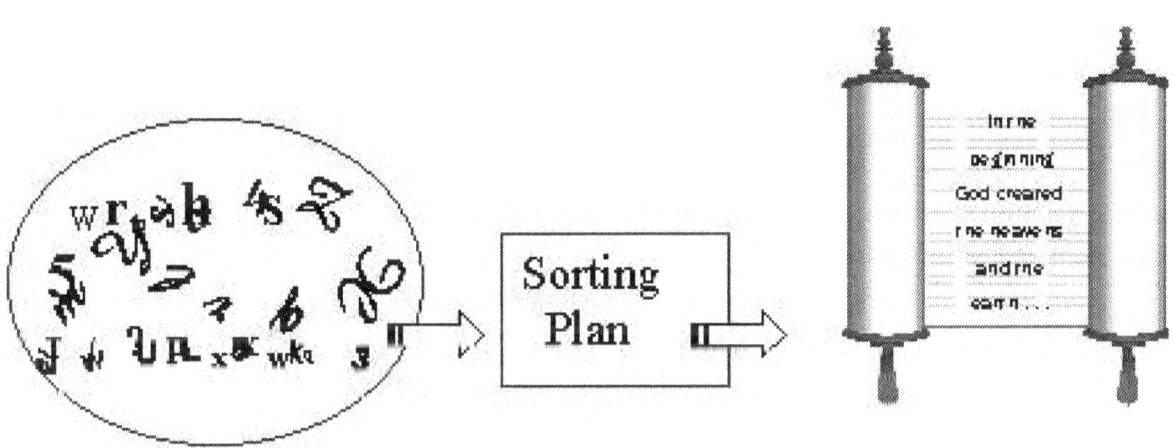

We can extrapolate this process for the entire Universe and the Earth in particular. If we had the entire Bible in a computer stored as a jumble of letters and punctuation marks, in order to sort them into a meaningful sequence we must have a definite program, a set of rules that would follow an overall plan.

The Four Fundamental Forces of the Universe:
God's Rules for organizing "chaos" into order

The Electromagnetic Force keeps atoms together and is the basis for all chemical reactions.

The Strong Nuclear Force binds the neutrons and protons together in the nucleus. This force is important in nuclear reactions like fission and fusion.

The Weak Nuclear Force determines radioactive decay, i.e., the spontaneous emission of alpha and beta particles from inside the nucleus.

The Gravitational Force is responsible for large-scale structure of the Universe, the formation of galaxies, stars, and planets.

The Anthropic Principle
"If any one of these forces is changed just slightly, the Universe would swallow itself up into a dark black hole or would collapse into a fantastically hot nugget."
Noble Laureate Steven Weinberg Ph.D., author of
<u>The First Three Minutes</u>
(Ref. 51)

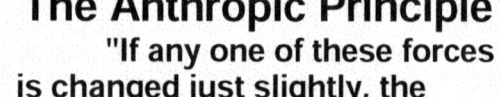

STRONG NUCLEAR FORCE

ELECTROMAGNETIC FORCE

WEAK NUCLEAR FORCE

GRAVITATIONAL FORCE

Creation of the Universe "Big Bang"

10^{-35} 10^{10}

10^{-43}

The Four Fundamental Forces of the Universe

Gravitational (attraction between masses): $F = \dfrac{GM_1 M^2}{1^{r2}}$

Particle: Graviton (Not observed)

Electromagnetic (holds atoms together, the basis of all chemical reactions):

$$F = \dfrac{e^2 1}{r^2}$$

Strong Nuclear (binds neutrons & protons together in the nucleus of the atom): $F = g^2 NN \dfrac{e\text{-}r/r^2}{\pi\lambda}$

Weak Nuclear (determines the radioactive decay of atomic nucleus): $F = g^2 \omega \dfrac{e^{-r}/\lambda\,\omega}{r^2}$

Particle: (not observed)

DAY 2-SKY / WATER

SUN

The disk of the Milky Way is formed from one of the spiral arms of our galaxy. A main sequence star, our Sun, is formed

And God said, "Let there be an expanse between the waters to separate water from water." So God made the expanse and separated the water under the expanse from the water above it. And it was so. God called the expanse "sky". And there was evening, and there was morning -- the second day". Genesis 1:6-8

DENSE WATER VAPOR

Milky Way and Sun not visible from Earth due to earthy atmosphere.

WATER

EARTH

DAY 2 equals 4 billion present day Earth years.

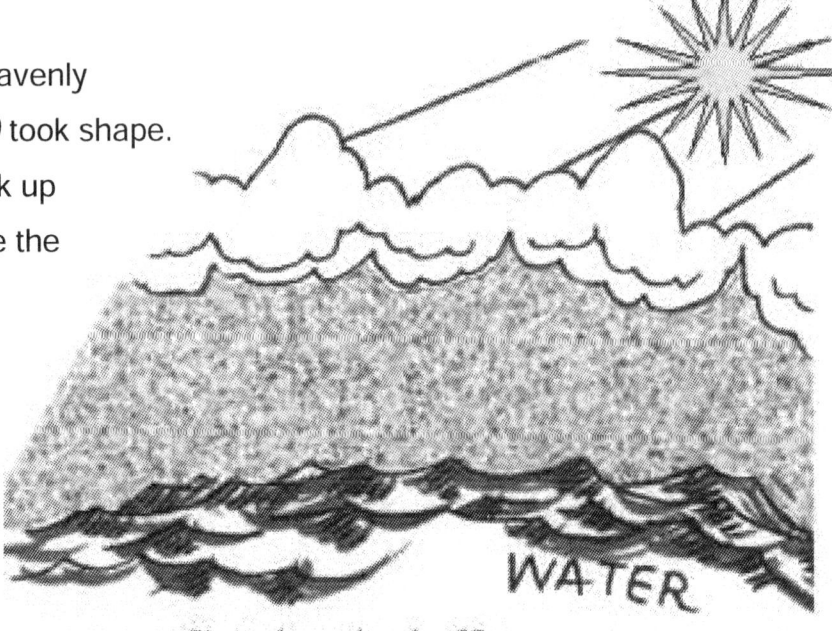

DAY 2 Sky and Water

The calculated span of DAY 2 is about 4 billion present day Earth years. During this period, the heavenly *Firmament* (meaning *sky)* took shape. (Genesis 1:8)   As we look up into the sky today, we see the sun (day time), the moon (night time), and beyond those we see the cloud-like spiral arm of our galaxy. We call this starry cloud the Milky Way, since it is composed of billions and billions of stars; however, if you had been standing on Earth between 7.75 and 3.75 billion years ago, you would NOT have been able to see any of these things because the newly formed atmosphere was too murky, something like the present atmosphere of Venus.

During the next day of Creation God would begin to clear the atmosphere when He creates the first plant life (algae) in the ocean. These plants will absorb the murky carbon dioxide and replace it with clear oxygen. Again we are reminded that God is changing chaos into order by the repeated Biblical statement regarding the evening (chaos), and the morning (order) of DAY 2.

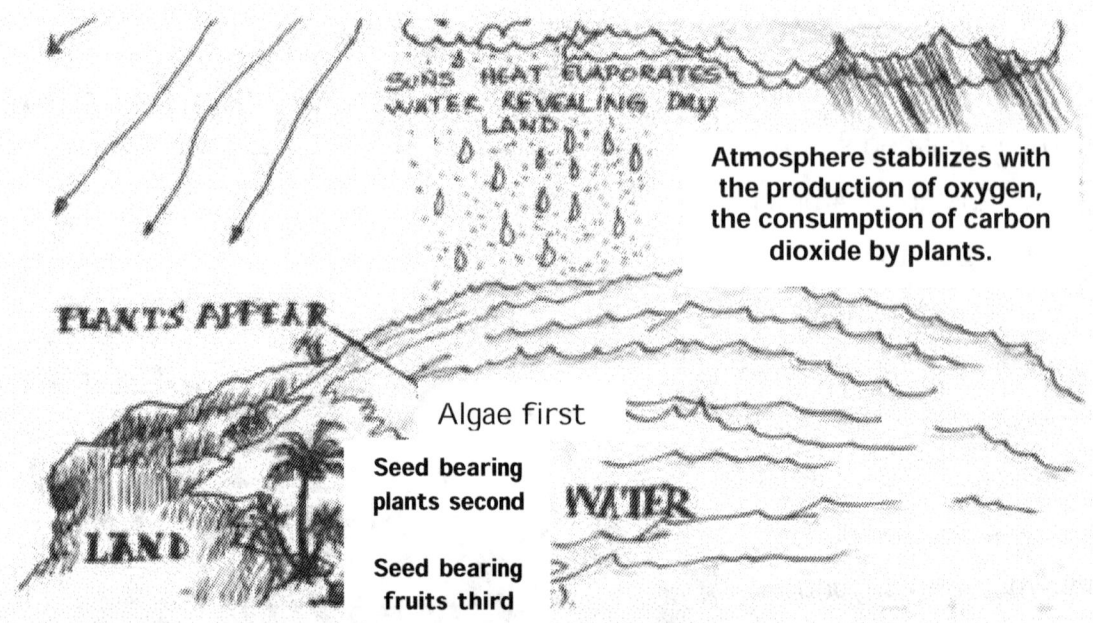

Atmosphere stabilizes with the production of oxygen, the consumption of carbon dioxide by plants.

PLANTS APPEAR

LAND

Algae first

Seed bearing plants second

WATER

Seed bearing fruits third

Day 3 of Genesis spans about 2 billion Earth years when we apply logarithmic time conversion formula. During this period is the appearance of dry land above the seas and creation of the first single cell plant life at 3.8 billion Earth years. Paleontologists have lately confirmed this Biblical time line. The fossils show explosive growth of single cell plant life (algae and bacteria) most immediately after the appearance of liquid water, just as the Bible says. At this point we have a caveat: Single cell algae and bacteria are not the same as the catalogue of vegetation including grasses, herbs, and fruit trees. True, this is the proper order of their appearance, but they were to come later . . . much later. Some flowering plants and trees were not found to have lived until about 120 million years ago. Nahmanides points out that there is no particular day of creation for all of plant life, since each one "is not a unique creation". (Ref. 31) Instead we simply find the beginning of plant life on DAY 3, with only the sequence of the more complex forms to be created later given here. Confirmation is found in recent DNA studies by molecular biologists that have discovered forms of single cell algae whose DNA contains more than 100 times as much genetic information as is found in that of the more complex mammals. (Ref. 8) Although created on DAY 3, the DNA instructions for the formation of more complex plants were activated later.

DAY 4 describes events
covering 1 billion Earth years.

Ultra violet light

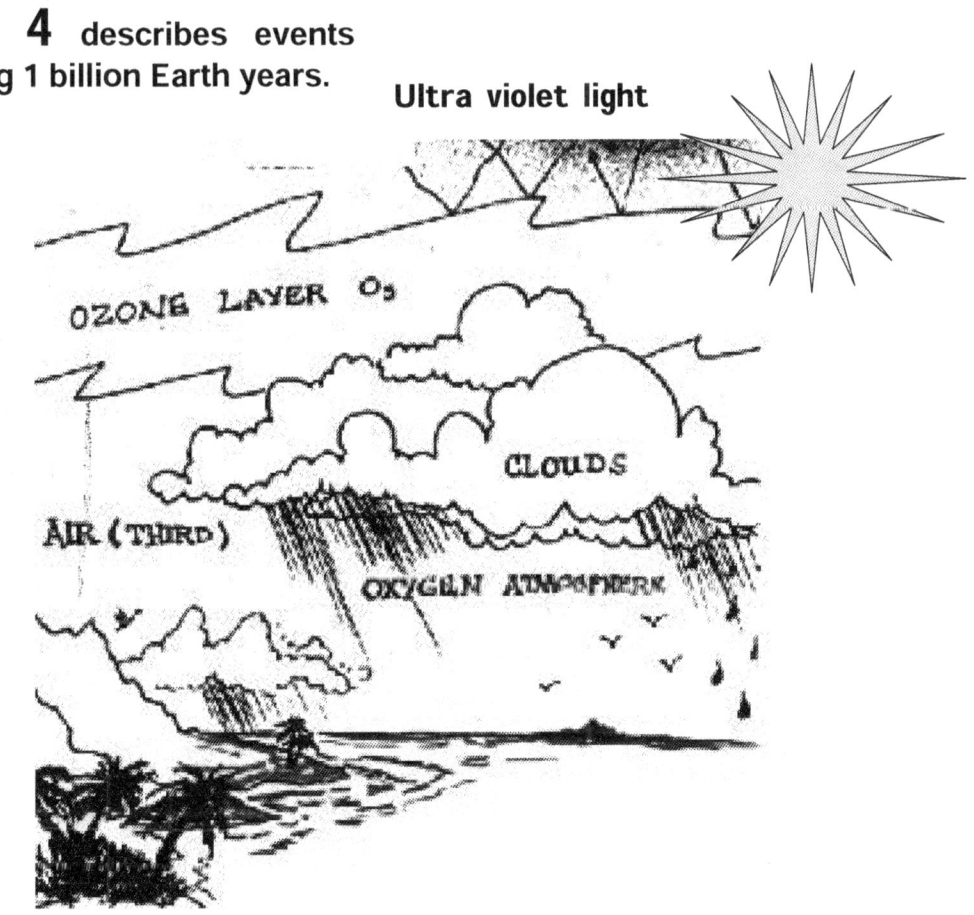

We remember that on DAY 2 the heavens were formed, but were not yet visible from the Earth. With the advent of plant life in the oceans (DAY 3), the process of photosynthesis begins as God creates chlorophyll, using plants to absorb CO_2 and release O_2 into the atmosphere

As more and more of the Sun's rays penetrate the overcast, the greater the production of oxygen until it reaches today's concentration, giving our present day's transparent atmosphere. Geologists tell us that this is exactly what the fossil record reveals. (Ref. 2) The Hebrew Talmud describes a similar scenario in its comments on these verses showing the divine revelation of God's plan of creation thousands of years (Earth) before mankind had the mathematics and insights to understand it. (Ref. 46) It is unfortunate that understanding and belief are NOT synonymous..

Day 4

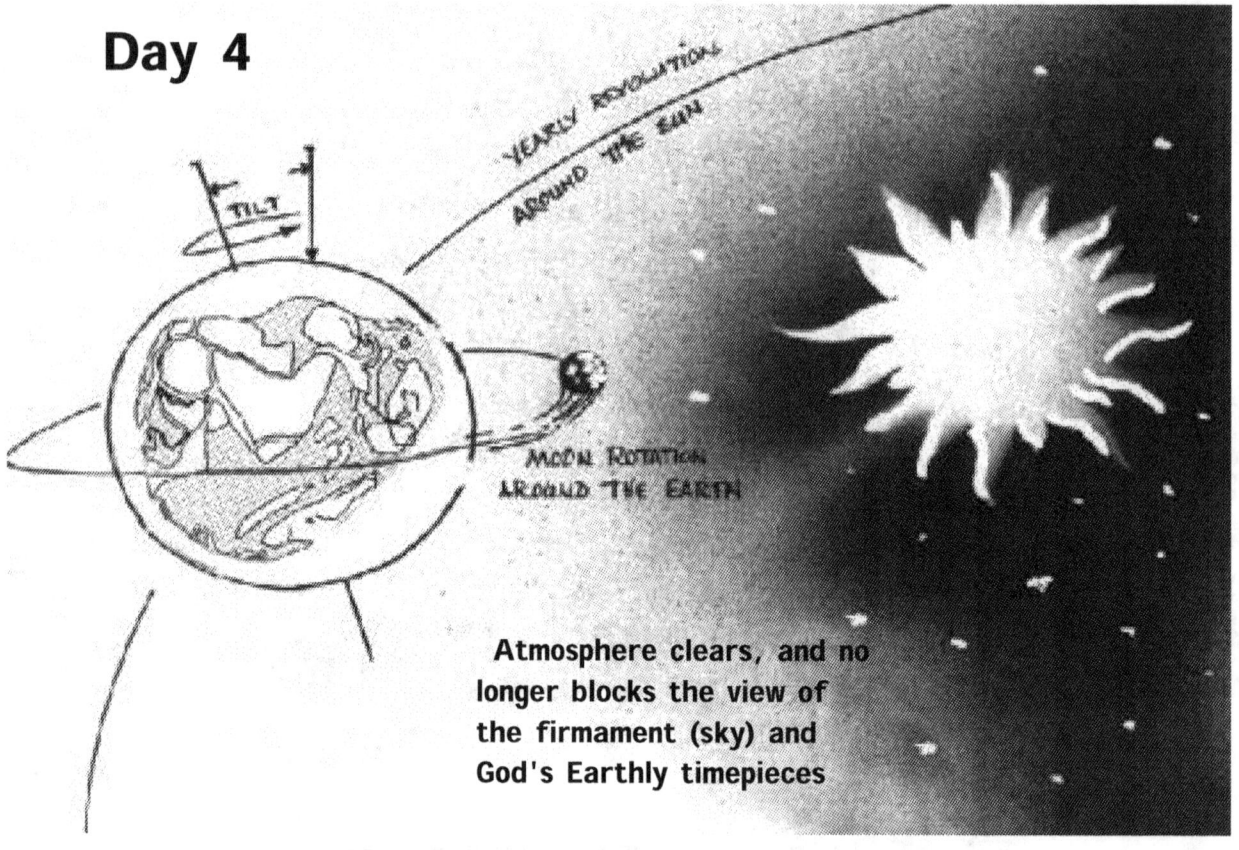

Atmosphere clears, and no longer blocks the view of the firmament (sky) and God's Earthly timepieces

> **Genesis 1 : 14 - 19** *"And God said, Let there be lights in the firmament of heaven to divide the day from the night; and let them be for signs, and for the seasons, and for days and years.*
>
> *And let them be for lights in the firmament of heaven to give light upon the earth' and it was so. And God made two great lights, the greater light to rule the day and the lesser light to rule and night, He made the stars also. And God set them in the firmament of heaven to give light upon the earth.*

Establishment of Earth's time scales
1. 24 Hour Day (speed of rotation)
2. Seasons (tilt on its axis of rotation)
3. Lunar Month (28 days)
4. Solar / Year (365 $^1/_4$) days

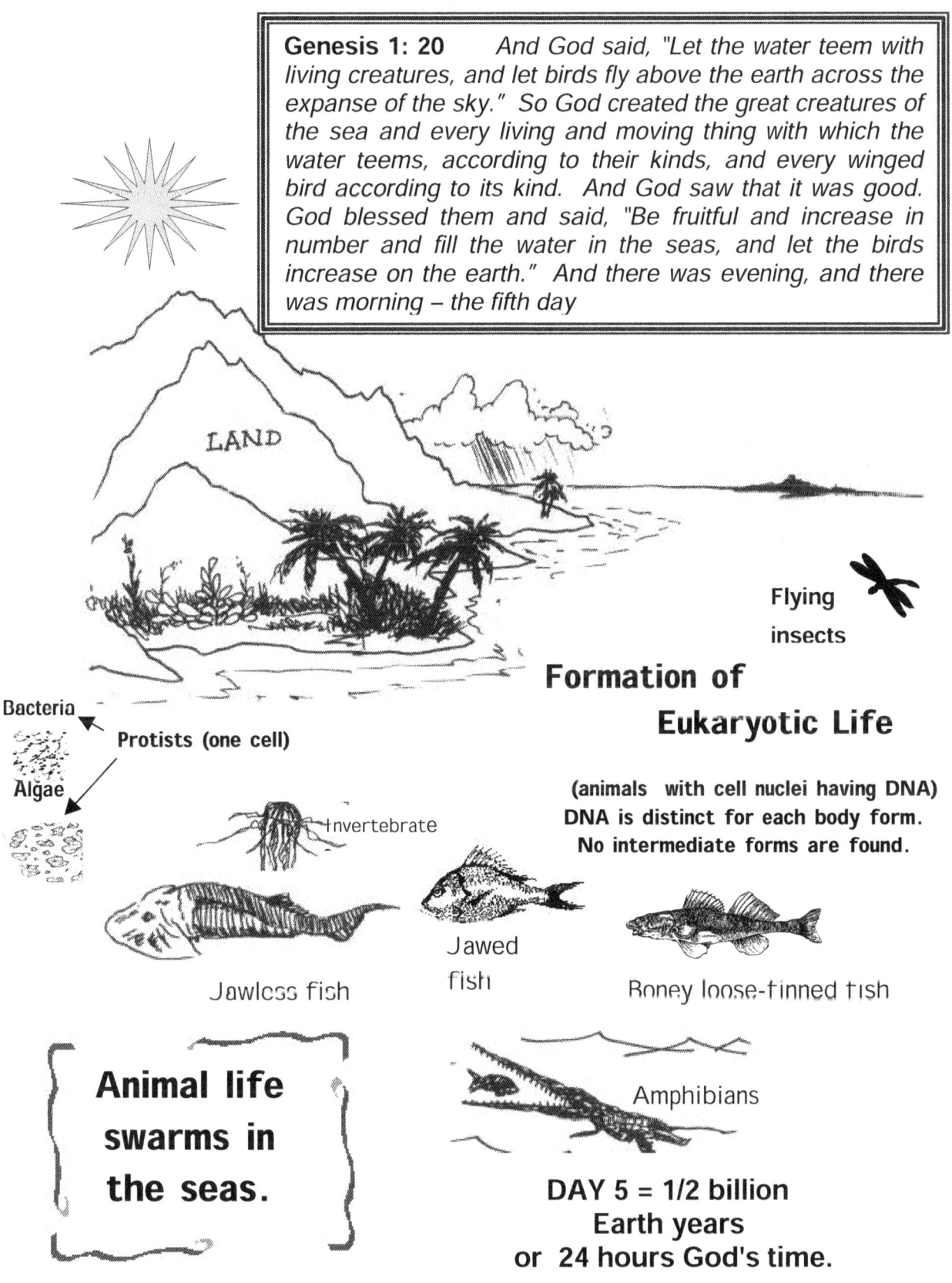

Genesis 1: 20   *And God said, "Let the water teem with living creatures, and let birds fly above the earth across the expanse of the sky." So God created the great creatures of the sea and every living and moving thing with which the water teems, according to their kinds, and every winged bird according to its kind. And God saw that it was good. God blessed them and said, "Be fruitful and increase in number and fill the water in the seas, and let the birds increase on the earth." And there was evening, and there was morning – the fifth day*

LAND

Flying
insects

**Formation of
Eukaryotic Life**

(animals with cell nuclei having DNA)
DNA is distinct for each body form.
No intermediate forms are found.

Bacteria

Protists (one cell)

Algae

Invertebrate

Jawless fish

Jawed fish

Boney loose-finned fish

**Animal life
swarms in
the seas.**

Amphibians

**DAY 5 = 1/2 billion
Earth years
or 24 hours God's time.**

DAY 5 Animal life occurs first in the seas as Genesis and the fossil record tell us. DAY 5 lasts about 1/2 billion Earth years, beginning approximately 750 million Earth years ago and lasting until 250 million Earth years ago. Beginning with single cell bacteria up through amphibian reptiles and flying insects, all basic body forms were designed into the original DNA. (Ref.16, 35). No, there is no evidence of "intermediate forms" appearing either in the fossil record or in the scriptural account. Sea life was so abundant that it exploded with many forms now extinct.(Ref.2)

Q What was the driving force behind this explosion of diverse animal life taking place after billions of Earth years of only single cell plant life?

A Many scientists believe that it was the increase in molecular oxygen by the respiration of algae that permitted multicellular animals to proliferate. This permitted a higher metabolic rate; thus organisms could the become more complex. Ref. 10)

We notice that the English translation of the Hebrew word "OAF" is "*bird*" (Genesis 1:20, N.I.V.); however, the Hebrew scholars tell us that a more accurate translation is "*winged animal*". The actual word for bird is "TSEPOOR", which occurs in Genesis 4:7. Therefore, the "*winged animals*" mentioned in Genesis 1:20 probably were the winged insects and/or reptiles that began life in the water and later flew above it. (Ref. 31)

DAY 6

Genesis 1:24-**31**

And God said, "Let the land produce living creatures according to their kinds: livestock. Creatures that move along the ground, and wild animals, each according to its kind." And it was so . . . And God said, "Let us make men in our likeness: and let them rule over the fish of the sea and the birds of the air, over the livestock, over all the earth . . ." So God created man in his own image, in the image of God he created him; male and female he created them. God blessed them . . . and the evening and the morning were the sixth day.

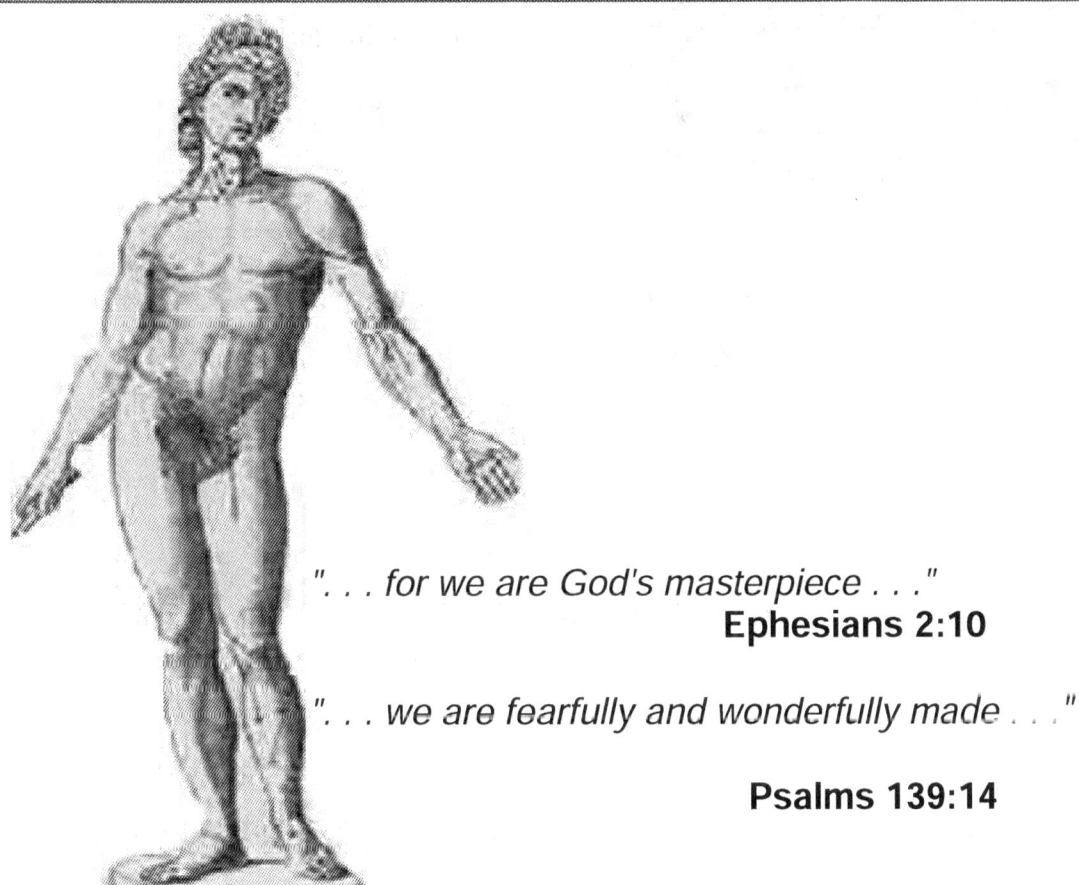

". . . for we are God's masterpiece . . ."
Ephesians 2:10

". . . we are fearfully and wonderfully made . . ."

Psalms 139:14

DAY 6

DAY 6 lasts 250 million Earth years, or 24 hours God's time, and is the shortest period of the Creation week. Yet during this period we see the creation of 34 present day phyla and 30 million species. Both the Bible and science agree that humans are the most complex and unique of all of the animals.

One of the species missing today is the dinosaur, yet they dominated animal life on Earth for 150 million Earth years.

Whatever happened to the dinosaurs?

This is a question that is posed to every parent by his or her wide-eyed children. This is not surprising, since these "thunder lizards" are a fascinating chapter of life that captures everyone's imagination. But the big question is:

Does the Bible mention Dinosaurs?

Yes, the Bible very clearly mentions dinosaurs!

Let us look at Genesis 1:21. We have already considered the creation of the DNA code for the 30 million plus species of animal life in DAY 5. Now we examine the specific list of animals presented in Genesis 1:21-24 - - - in Hebrew, not in English. Hebrew scholar Dr. Gerald Schroeder explains that among this list of animals mentioned is one classification named **"*TANINIM GEDOLIM*"**.

There is no problem with **GEDOLIN** because it is the Hebrew word for *BIG*, but **TANINIM** has a variety of meanings, some of which are:

**Sea
Monster**

Dragon

Alligator

Which fits the best? Dr. Schroeder points out that **TONINUM** is the plural form of **TANEEN** (singular) that appears in Exodus 7:10. Here we are told that when Aaron casts his rod down before Pharaoh, it becomes a **TANEEN**, not **NAHASH** (Hebrew word for snake). **TANEEN** is the general classification to which NAHASH belongs: reptiles. (Ref. 34)

God knew, and now we know
what those big reptiles were.

The Dinosaurs were biggest reptiles that ever lived. They may have been unknown to Moses, but he faithfully wrote down what God said.

Paul Garnett

The reign of the dinosaurs came to an abrupt end about 65 million Earth years after dominating animal life on the Earth for 150 Earth million years. What happened?

Picture a 7 ton, 30 foot long brontosaurus and several other prehistoric reptiles in an ancient swamp located on what is now the plain of central North America.

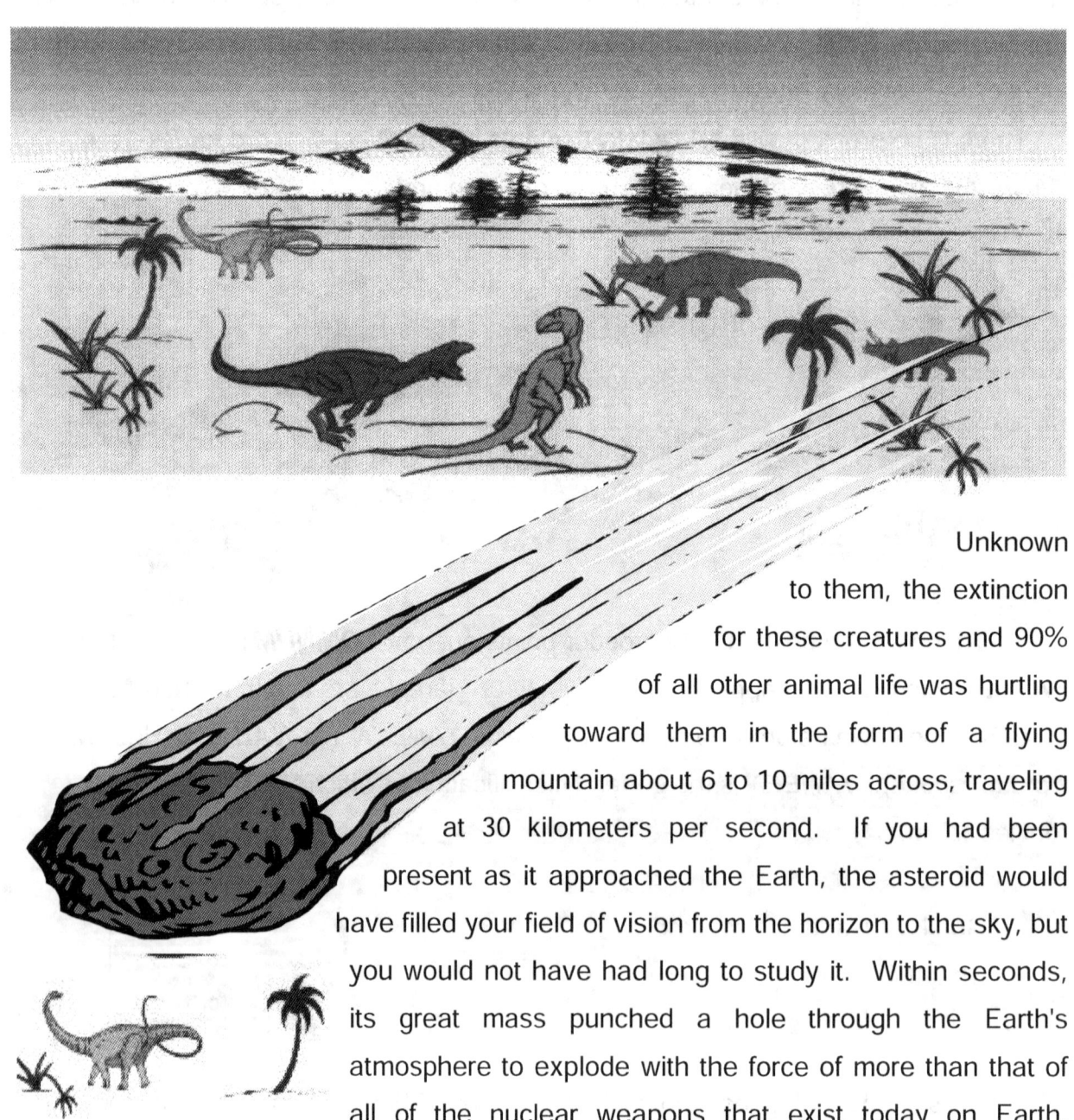

Unknown to them, the extinction for these creatures and 90% of all other animal life was hurtling toward them in the form of a flying mountain about 6 to 10 miles across, traveling at 30 kilometers per second. If you had been present as it approached the Earth, the asteroid would have filled your field of vision from the horizon to the sky, but you would not have had long to study it. Within seconds, its great mass punched a hole through the Earth's atmosphere to explode with the force of more than that of all of the nuclear weapons that exist today on Earth.

The shock wave shook the entire globe. Volcanic eruptions burst forth along the continental tectonic rift lines. Although the point of impact was in the Caribbean, on the opposite side of the Earth in what is now India, great fissures opened up on that continent pouring molten lava to a depth of one kilometer in thickness.

Volcanic dust and gasses shrouded the Sun for months. In the darkness and intense cold, most of the plant and animal life died . . . and so did the dinosaurs.

One of the largest animals to survive that long cold night was a species of small mammal that probably looked much like a rat or squirrel.

Life on Earth had been redirected from cold-blooded, huge, stupid reptiles to warm-blooded small little mammals that would eventually dominate the world as they do today. God made a new start that would conclude with the creation of human kind, who would have a soul and be able to worship the Creator.

Is it possible for such a disaster to reoccur today? Astronomers tell us that it is not IF, but WHEN. The Bible speaks of this in Mark 13-15, where it describes "the stars falling from the skies"; however, our discussion here is of Genesis, not Mark.

The creation of all animal life on Earth is directed toward a single purpose: the creation of mankind. What about early man-like hominids?

Who were the Neanderthals?

If you were to encounter a Cro-Magnon today on the street, would you recognize him immediately as a "caveman"?   Were there "men" on Earth before Adam? Was Adam really the "first" human?

To avoid these questions, we could take the foolish attitude that such creatures never existed, or that they are the constructs of misleading paleontologists, despite the preponderance of fossil evidence to the contrary that is in existence.

A much better approach is for us to study carefully exactly what the original word of God says about them in the Bible. It must be remembered that Moses probably knew nothing about these man-like creatures, so their existence would be revealed by insight into the Genesis record given him by God. What we find in this examination will be astonishing, since Biblical truth will bring us right up to our present day relationship with the Almighty.

The key to understanding this is based upon the translation of two Hebrew words:

NESHAMA   meaning: the "*SPIRITUAL SOUL*" of human beings   (Genesis 2:7)   (Ref 32)

NEFRESH   meaning: the "*NON-SPIRITUAL ANIMAL SOUL*" (Genesis 1:21)   (Ref. 31)

> ". . . the Lord God breathed into his (Adam's) nostrils the breath of life; and man became a living <u>soul</u>." Genesis 2:7

It is the **NESHAMA** that makes Adam --- and his descendants --- different from all other life on Earth. Unlike all other creatures, God gave Adam (and Eve) a soul, NOT present in Cro-Magnon or previous hominid types. (Ref. 32)

The difference between Adam and Neanderthal or Cro-Magnon was not that Adam had a great super-brain. Comparing the three, we discover that their brains are about the same dimension: about 1.4 liters. (Ref. 55)

They both wore the same size hats!

No, it isn't the brain size or the body hair that separated Adam from previous hominids; it is something only mankind has: it is the **SOUL** of the human, or spirituality, as the Hebrew language tells us. When God said, ". . . let us make man in our own image . . ." (Genesis 1:26) does that mean we physically look like God? Perhaps, but more likely God meant that humans were given a soul . . . a spirit . . . like God who is **THE SPIRIT**. (Ref. 32) Let us not become confused with creatures that had similar features to Adam, but who appeared before God created Adam. Neanderthal and Cro-Magnon possessed **NAFRESH**, the animal soul (Ref. 32), and were part of the animal kingdom but were not humans (Homo Sapiens).

 The Hebrew word **NEPHILUM** means: "*less than*" or "*inferior*". (Ref. 33) Adam probably realized that the Cro-Magnon people were "less than" humans who possess souls. Yes, it is evident that they were still around for Adam to be able to make this comparison.

The moment God "breathes the breath of life" into Adam and instills in him a living soul (**NESHAMA**) on DAY 6, Earth's 24 hour day begins to tick and man begins to gauge time by the Earth's position in relation to our Sun. The Bible sets aside the exponential time flowing Cosmic Clock, and from this time forward Adam and his descendants keep linear 24 hour time (as measured between the two locations of the Sun and the Earth).

If Adam is a new and different being having a divine soul, we should be able to find some evidence of this change from animal behavior to a human personality. Of course, the gift of a soul leaves no physical evidence for anthropologists to measure. Let us consider the aspects that are possible to evaluate, such as the changes of behavior as evidenced from:

    1) transmission of information (writing),
    2) the better utilization of resources (metallurgy, farming, wheel),
    3) social organization (cities),
    4) an awareness of a Divine Creator.

Evidence of all these changes must occur in Mesopotamia (Adam's home), not in some distant continent such as China or Africa. This evidence should show that Adam's descendants are markedly different from the hominids that preceded him.

Let us consider the following aspects of "human-like" behavior and determine when and where they occurred.

#1   *WRITING:* transmission of information from one generation to the next

Writing first appears about 5500 BP (before present). Adam was created about 8000 BP.

These are the elements of the first communications mankind ever recorded.   Information could now be transmitted to future generations, and it was no longer necessary to reinvent useful tools or ideas because they could be described and preserved. Records could now be kept of ownership and of laws. In addition, the use of numbers appeared in the exact same location, Mesopotamia. This was largely due to another human stride forward: the unique social institution, the City. (Ref. 20)

Meaning	Outline Character 3000 BC	Cuneiform about 2000 BC	Assyrian about 700 BC	Baby-lonian about 500 BC
The Sun				
God or Heaven				
Mountain				
Man				
Ox				
Fish				

32

.#2: Better utilization of resources
(metallurgy, farming, wheel)

Two of the many advances in technology that have been significant to the human race are:

The wheel:
considered by some to be the most important advance that humans ever produced.

Sledge and wheels attached to axle

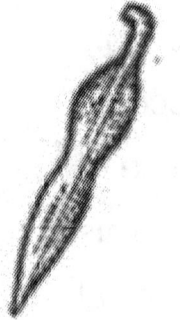

Refinement of metals
(copper tools, weapons and ornaments)

**By mixing in a little tin, Adam's descendants achieved a golden hard metal: BRONZE.
The Bronze Age had begun. (Ref. 55)**

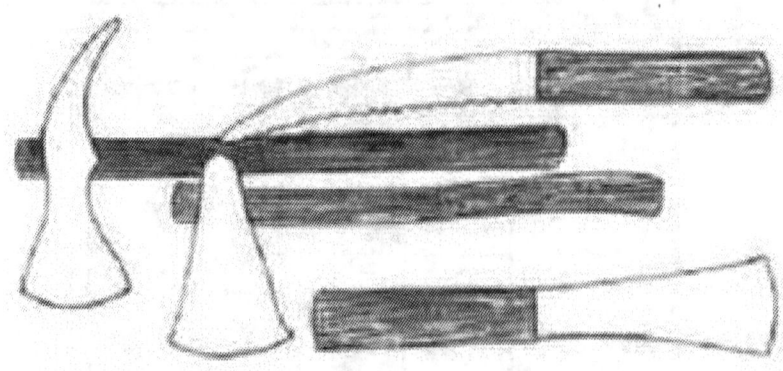

Bronze age carpenter tools

#3   DEVELOPMENT OF CITIES

The single greatest difference between Cro-Magnon and Adam's descendants is the change from nomadic hunter-gatherer life to village agriculture.

Cro-Magnon cave drawing showing hunters killing deer

Sketch based on a bas-relief of early Mesopotamian farmer cultivating wheat

**Notice
the difference in art.**

In Genesis 1:30 we read:

And God said, *"I give you (Adam) every seed bearing plant . . . and every green plant for food."*

Cro-Magnon and his predecessors had wandered the Earth for thousands of years, and now there is a sudden cultural change . What happened? Paleobotanists have found evidence of a burst of new vegetation after the last Ice Age. God not only gave Adam a soul and promised him all of the animals, but He also promised "every seed bearing plant" as food.

One of the new plants that God gave Adam was to transform mankind forever. It was a new hybrid wheat, unknown on Earth until this time (Ref. 26), that suddenly appears by a miracle of genetics. It is a wild wheat, which looked much like wild grass, crossed (combined with) natural goat grass to form a new hybrid that is not only a much bigger grain, but unlike most hybrids, is fertile (can be planted to raise similar plants).

This genetic "miracle happened when the 14 chromosomes of goat grass combined with the 14 chromosomes of wild wheat to produce 28 chromosomes of the new enlarged grain known as EMMER.

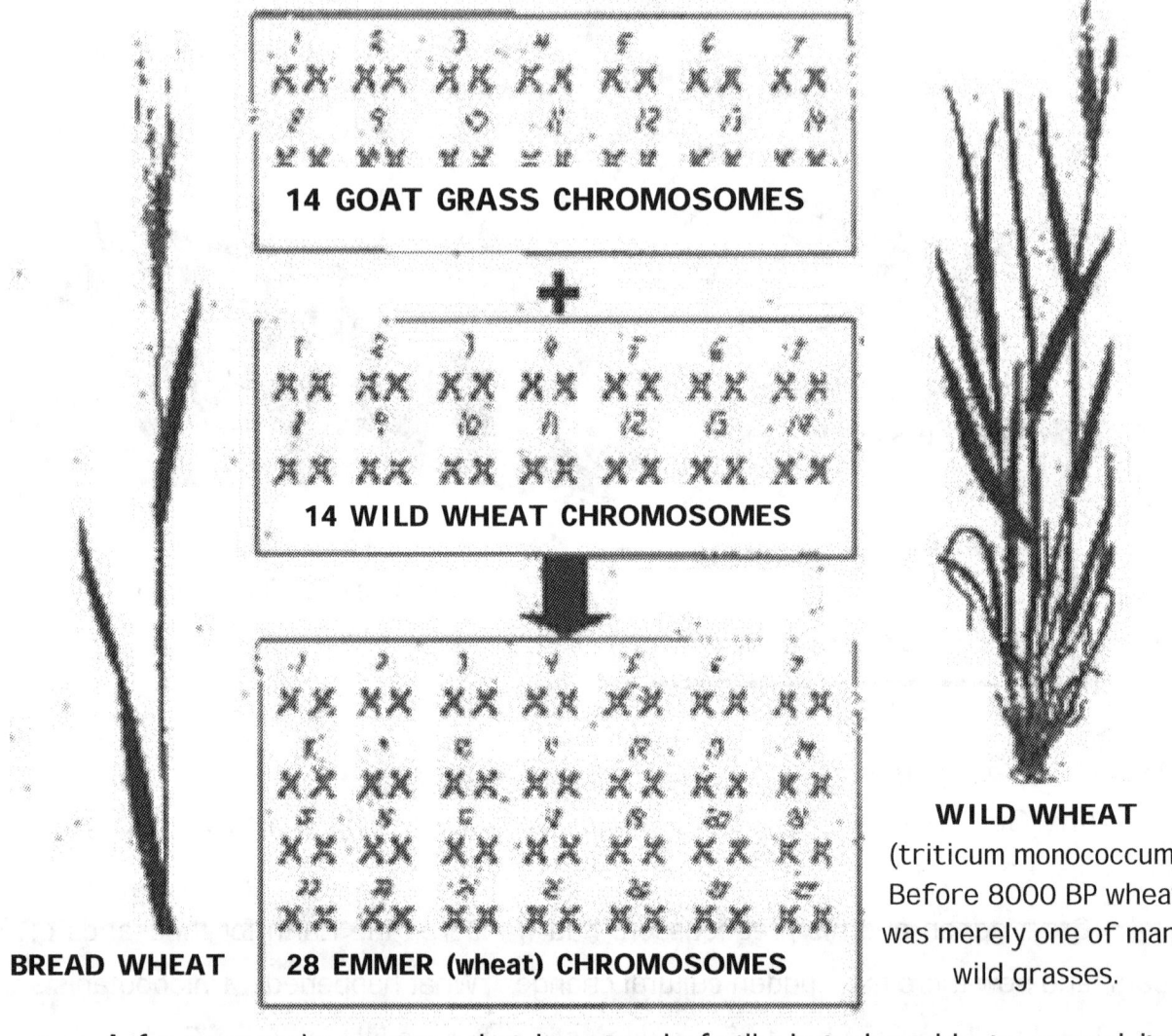

14 GOAT GRASS CHROMOSOMES

14 WILD WHEAT CHROMOSOMES

BREAD WHEAT **28 EMMER (wheat) CHROMOSOMES**

WILD WHEAT
(triticum monococcum)
Before 8000 BP wheat was merely one of many wild grasses.

A fat new grain emerges that is not only fertile but also able to spread itself naturally. The miracle does not end there. As EMMER spreads across Mesopotamia, it crosses with another goat grass (14 chromosomes) to produce an even larger hardier fertile hybrid (48 chromosomes). We know this grain today as BREAD WHEAT (Ref: 9.1).

God had created the "plants for food", and thus opened the door for civilization, since Bread Wheat must be planted, cultivated, and harvested by permanently established mankind, not wandering nomads. Thus cities began springing up in Mesopotamia, Adam's homeland.

4   Awareness of the Divine

Adam's descendants show something that no animal displays: they are supremely conscious. They are conscious of their own existence, conscious of the beauty and order of nature, and most important, conscious of the Divine Personality. All humans everywhere have Gods of some kind.

Virtually all humans believe in life after death. If God communicated directly with Adam, it is not unusual that his descendants should inherit not only the NESHAMA soul, but also an awareness of the Divine.

The Bible gives us a continuing record of God's communication with humanity through his prophets and his Son, Jesus of Nazareth. When we study these sayings, we discover the God-given fundamental concepts of all human behavior.

Some of these are:

➤ Love is the foundation of all justice (1 Corinthians 13)

➤ Human life is sacred   (Matthew 10:30)

➤ The importance of the dignity of the individual

        (Luke 12:6-7)

➤ Each individual has responsibility to the community.

        (Mark 12:17)

➤ All individuals are equal before the law (Romans 3:23)

➤ Peace is the optimum human condition (1 Peter 3:10-11)

The above references are just a sampling of the concepts that occur throughout the Bible in both Old and New Testaments.

(Suggested by Paul Johnson)

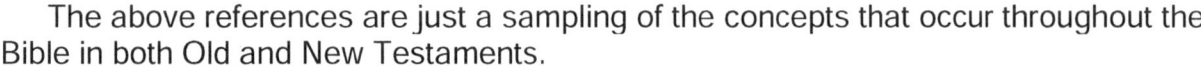

The descendants of Adam were:

a. **Inventive** (created wheel)

b. **Communicative** (developed writing)

c. **Socially aware** (developed cities)

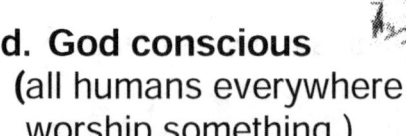

d. **God conscious** (all humans everywhere worship something.)

All these advancements must occur in one place, Adam's homeland, Mesopotamia, not some in China, some in Europe, some in Africa, but all together during one period of time as Adam's descendent populated Mesopotamia. This is exactly what the archeological record shows.

Animals are not aware of beauty or the symmetry of nature.

Animals are focused on food, survival and sexual mating. (Ref. 6)

Adam and his progeny are profoundly different because humans have an everlasting soul and are aware of God, themselves and the beauty of Creation. (Ref. 45)

"Thus the heavens were completed in all their vast array. By the seventh day God had finished the work he had been doing; so on the seventh day he rested from all his work. And God blessed the seventh day and made it holy because on it he rested from all the work that he had done." **Genesis 2 : 1-3**

DAY 7

There is far more to Creation than the sweeping statements of Genesis. There is beauty, harmony, grace, balance, and purpose. Yet the God of Creation did not require 7 days, be they billions of Earth years or Pico/seconds, to produce the Universe. He took whatever time He wished; He did so because it pleased Him. (Genesis 1:31) It also pleased Him to leave us a record of those 6 days, whatever their duration, to be studied 3,300 years later. Of all the ancient human accounts of creation, only Genesis makes enough sense to be verified by the various disciplines of science.

Actually, we don't need some external "proof" of God's creation, since God Himself tells us in His record that all mankind has been given 2 proofs of His Deity:

1) The internal voice of our conscience that speaks to us each day as we bend or break the laws He has given us to live (or die) by (Romans 2:15).

2) The beauty, scope and sequence of the sparkling night sky. (Psalms 10:1)

A
Short
History
of
Time

According to:

Newton

Einstein

Sir Isaac Newton developed his laws of motion and gravity by building on the work of others. His theory of Gravitation provided a model which man's mind could grasp. It made sense out of countless phenomena, and could be tested and measured in the laboratory.

In Newton, science at last seemed to have freed itself from the elusive dependence upon philosophy. Many felt Newton had provided the means for intellectual liberation and now they could escape from all that was blindly conventional.

Newton's Law of Gravitation says:

"Objects are attracted to each other by the amount of mass (matter) they contain, but inversely by the square of the distance between them."

Gravitation, you will recall, is one of the Four Fundamental Forces of the Universe. It keeps the planets, stars, and galaxies balanced in space over vast distances, and explains (almost) their motion. We say "almost" because it did not quite account for the motion of the planet Mercury. This made Einstein suspicious that there was more to gravity than Newton had supposed.

Einstein postulated:

I have two proposals here. Watch closely, please.

1. NOTHING CAN TRAVEL FASTER THAN LIGHT.
(All observations of the speed of light are shown to be equal.)

2. **ALL SPEED, MOTION, & TIME ARE RELATIVE TO THE OBSERVER.** (Because there is no observer's platform that is not also in motion): This is known as Relativity

Galileo Galilei

Einstein was now ready to deal with an important factor, which Newton did not explain.

Galileo showed that a lead ball and a feather ball of the same size, when dropped from the Leaning Tower of Pisa, would HIT THE GROUND AT THE SAME TIME!

One would expect the lead ball to hit first. To Einstein this meant that their falling speed had NOTHING TO DO WITH WHAT THEY WERE MADE OF (ignoring air resistance). Maybe there was a another way of explaining gravity. May be gravity was not a force at all. Maybe the effects attributed to gravity were actually due to some special characteristics of space.

Why should Mercury have a different orbit from other Planets?

It took a young Jewish patent clerk working in Switzerland to challenge the monumental edifice of Newton's Gravity, a theory that was so precise it could predict the existence of the planet Neptune even before telescopes were developed to discover it. Einstein began by discarding two of Newton's fundamental concepts.

1 Time CANNOT be the same everywhere in the Universe.

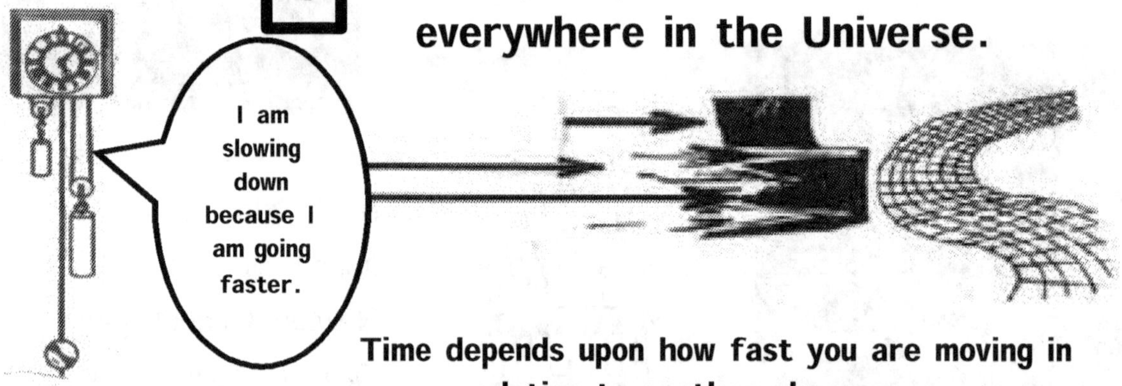

I am slowing down because I am going faster.

Time depends upon how fast you are moving in relation to another observer.

2 Space IS NOT the same everywhere.

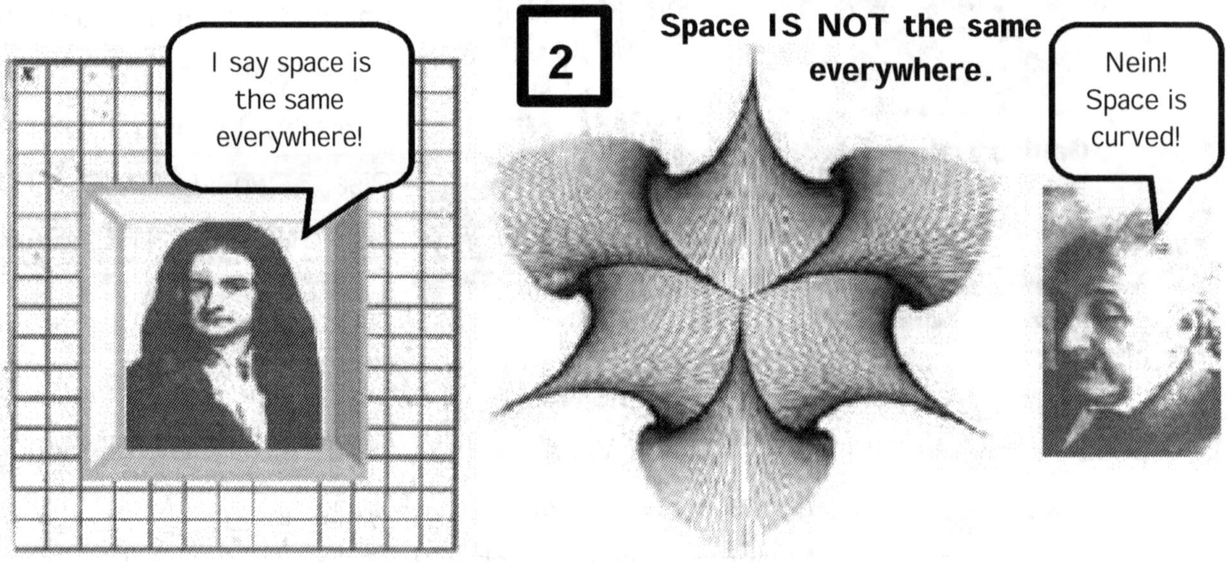

I say space is the same everywhere!

Nein! Space is curved!

Is space flat or curved? Einstein suspected it is curved because of the curved orbits of the planets; but how could he prove this mathematically? He went to his friend Marcel Gossmen and insisted that Gossmen teach him the special mathematics of curved Riemann Geometry to help him create the equations of curved space. Working with these equations led Einstein to conclude that space IS indeed curved, and the local curvature is produced by the presence of mass.

Here is what Einstein called a "Gedanken", or "thought experiment". See if it works for you. Think of a frame with tight rubber sheet stretched over it. Now roll a ping-pong ball across it. Will it travel in a straight line? Yes. This demonstrates an object passing through FLAT space without other mass in it.

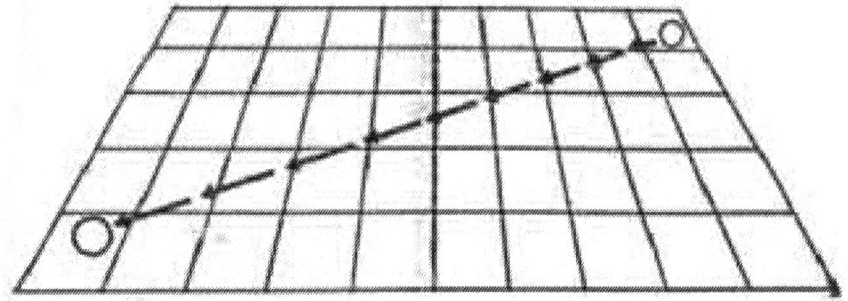

Let us now place a heavy bowling ball in the middle of the sheet, causing a depression in the center. We see a model of how space is curved or distorted by mass. The heavier the mass, the more distortion there is.

MATTER TELLS SPACE HOW TO CURVE,

AND THEN SPACE TELLS MATTER HOW TO MOVE

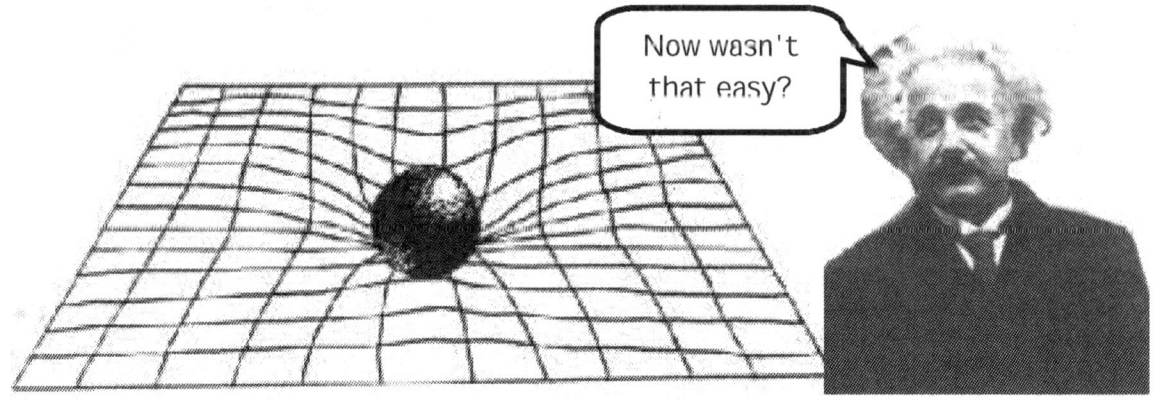

In most cases, the distorted pattern of curved space, caused by a heavy mass capturing a moving object, follows the "Geodesic" lines of curved space and traces out a circular orbit.

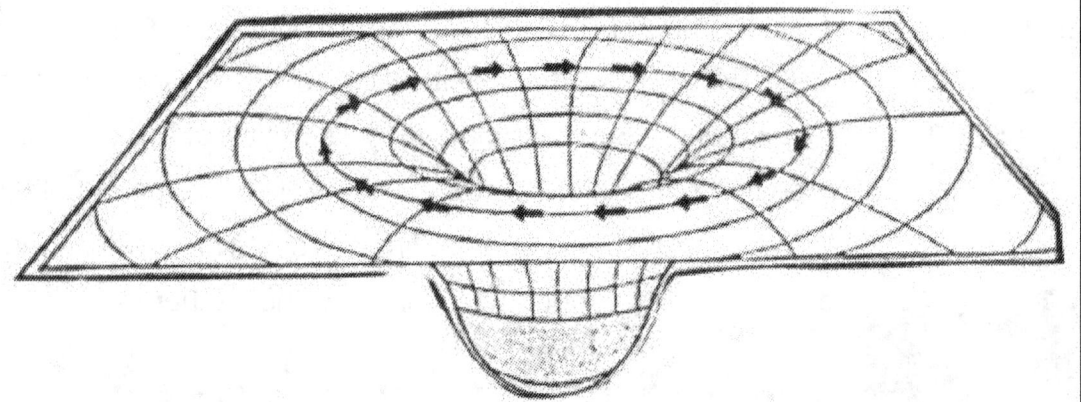

Occasionally, when the speed of the moving object is great enough and its path is headed directly towards the center of the distortion caused by the mass, it will accelerate, crashing into the mass, such as a meteorite crashing into the Earth. Notice that its path is still curved, however.

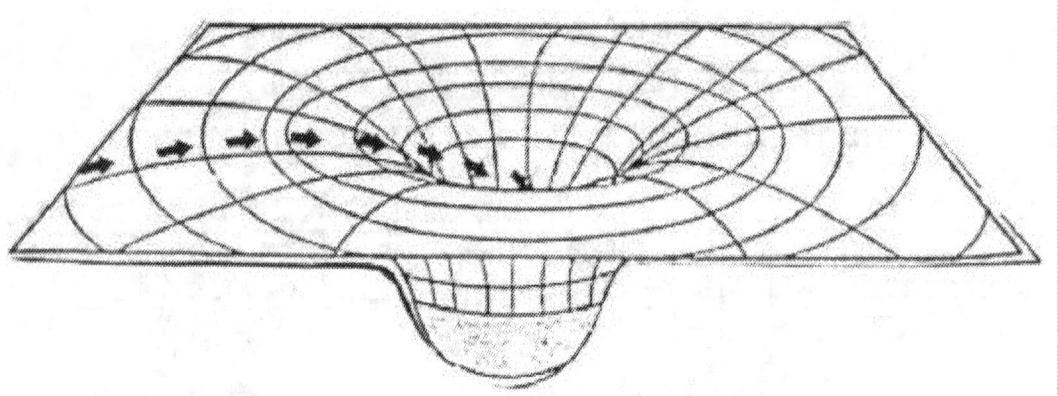

The quiet little eccentric scientist sitting in his study doing Gedanken (thought) experiments had come upon a part of the great plan of the Almighty for the whole Universe. Now he must prove it.

The first real test of Einstein's proposal came with a total eclipse of the sun on May 29, 1919. He predicted that starlight passing just at the edge of the sun would be displaced by 1.7 seconds of arc from its actual true position. An English astronomer, Sir Arthur Eddinqton, was chosen to lead..

OK, so you are tired of doing Gedanken? Let's do a REAL experiment. Let's use the whole Universe and see what happens .

expedition to the small island of Pricipe, off the coast of west Africa, to photograph the eclipse to see if there really was the displacement that Einstein had predicted. The results amazed the scientific world. Eddingon's photo plates showed that light rays that had left distant stars thousands of years ago were bent by curved space near the sun only 8 minutes before they marked their image on his plates. This was the exact amount predicted by Einstein. "Gedanken" had been confirmed by reality. Einstein became world famous overnight

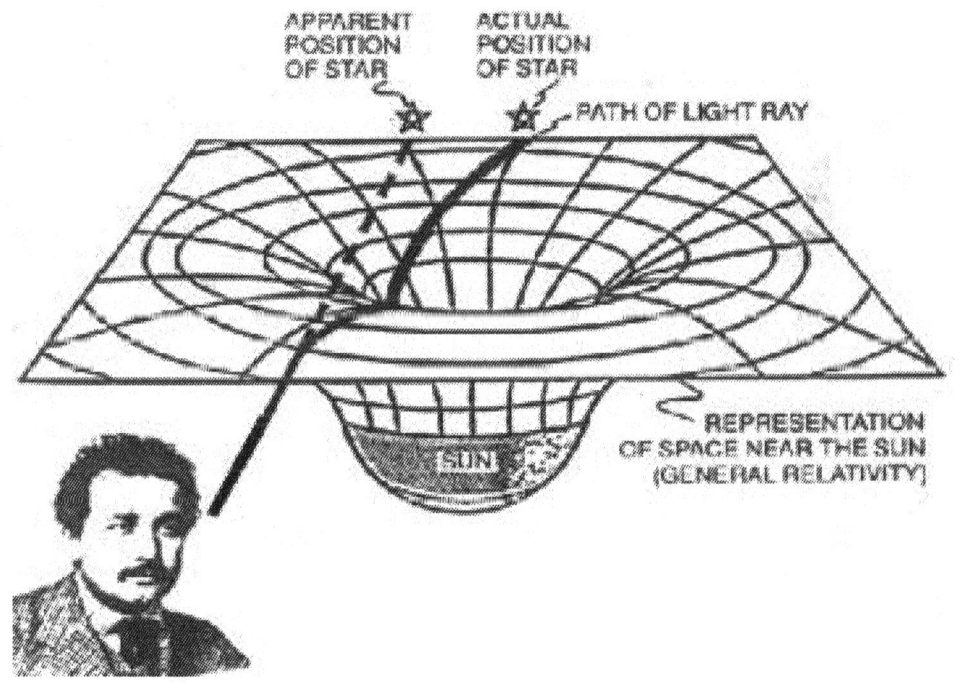

APPARENT POSITION OF STAR

ACTUAL POSITION OF STAR

PATH OF LIGHT RAY

SUN

REPRESENTATION OF SPACE NEAR THE SUN (GENERAL RELATIVITY)

The dramatic proof of Einstein's equations was to have a profound effect upon mankind. In 1938, a German scientist, Hans Bethe, predicted from Einstein's equation, $E=MC^2$, that atomic fission would release an enormous amount of energy from within the atom. In 1939, Germans, Otto Hahn and Fritz Stassman, discovered nuclear fission, and Swedish Niels Bohr published a paper on how a nuclear bomb could be constructed.

Fortunately, the scientifically ignorant Hitler took no notice of these findings, but Einstein

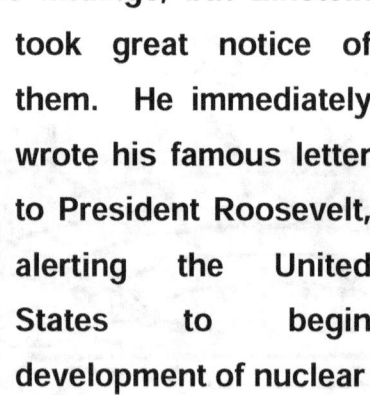

took great notice of them. He immediately wrote his famous letter to President Roosevelt, alerting the United States to begin development of nuclear energy.

Was it by chance that Hitler's cruelty drove out of Germany the very person who could have given him the power to rule the world? Think of what a dark curtain would have descended over civilization if Hitler had gained the atomic bomb first. No, not by chance. God controls both the physical and human destiny of this world. —

T.O.E.: The Theory of Everything

T.O.E. explains all the wonderful parts and happenings in the Universe and how the Universe itself began. It tells us about the frantic dance of subatomic quarks to the majestic swirl of the distant galaxies. Not only all of these things, but also the electromagnetic nerve impulses flashing through your brain right at this moment.

 What is T.O.E. all about? Built on Einstein's equations, J.J.Thomson, Ernest Rutherford, Niels Bohr and James Chadwick in the 1930s established the atomic model, which looked like a miniature solar system as the ultimate description of all matter.

But in 1968, atomic scientists at Stanford Linear Accelerator Center found that protons and neutrons were not the ultimate ingredients of matter, but that they consist of three smaller particles called "quarks" (a poetic term from James Joyce's *Finnigan's Wake*.). Furthermore, the quarks come in two classes, unimaginatively named "up" and "down" quarks.

Proton	Neutron
quarks:	quarks:
up	down
up	down
down	up

All objects to be seen here on Earth and in the heavens are made up of electrons, and up and down quarks. There are many other particles but these are largely the results of bombarding the atomic nucleus with high-energy atom smashers. Scientists have probed the structure of matter down to about a billionth of a billionth of a meter to find these ultimate ingredients of matter.

To help us understand the components of matter, let us consider them.

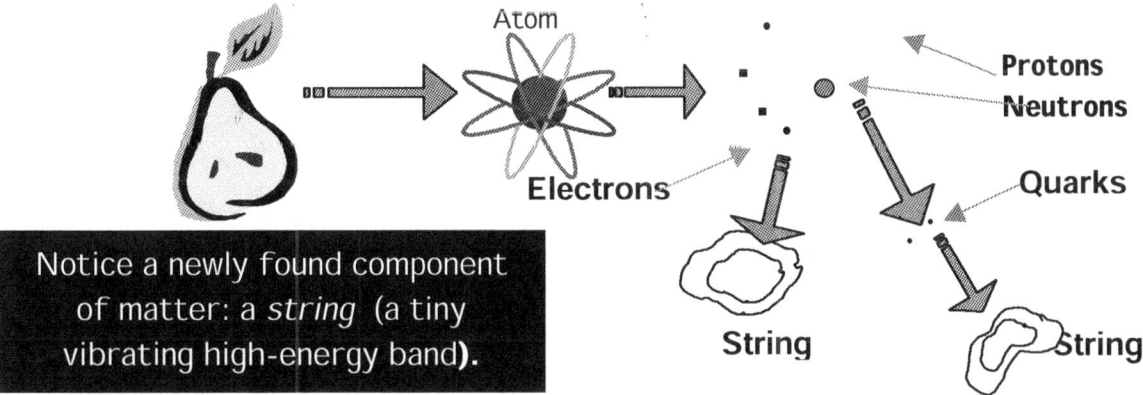

Notice a newly found component of matter: a *string* (a tiny vibrating high-energy band).

String theory has replaced the point-like particle with a tiny one-dimensional high-energy loop so small it is unlikely that even our best instruments will ever reveal it. A string loop is about 100 billion billion (10^{20}) times smaller than an atomic nucleus. The theory predicts that each string vibrates at a different resonant frequency, which determines the physical characteristics of the particle. An electron is a string vibrating one way and the up quark is a string vibrating in a different way; consequently, particle properties (mass, charge, etc.) are simply resonant patterns of the fundamental   vibration of the string.

Picture a guitar string. When plucked at its fundamental frequency, the string vibrates:

This pattern depends upon:

1) *Amplitude*, (*the displacement* between peaks and troughs), and

2) *wave length* (the separation between one peak and the next).

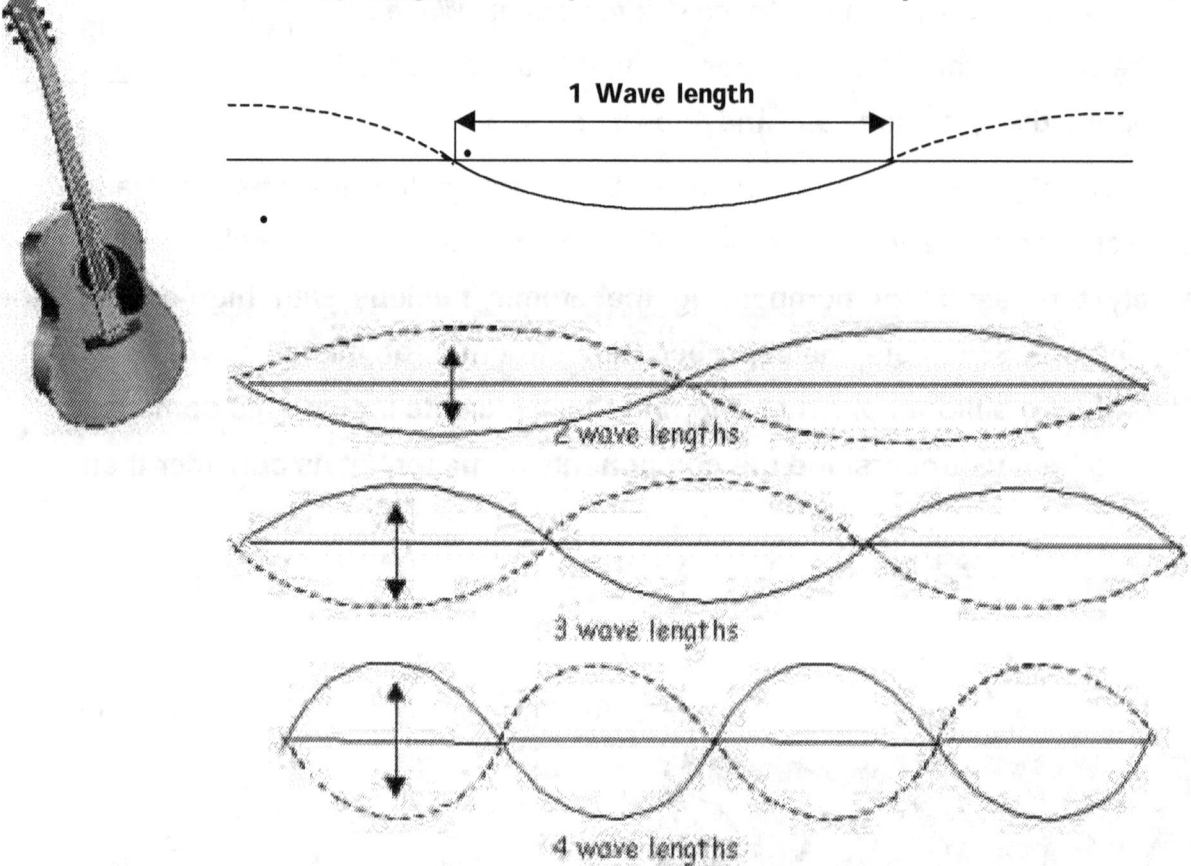

1 Wave length

2 wave lengths

3 wave lengths

4 wave lengths

The same is true of loops.

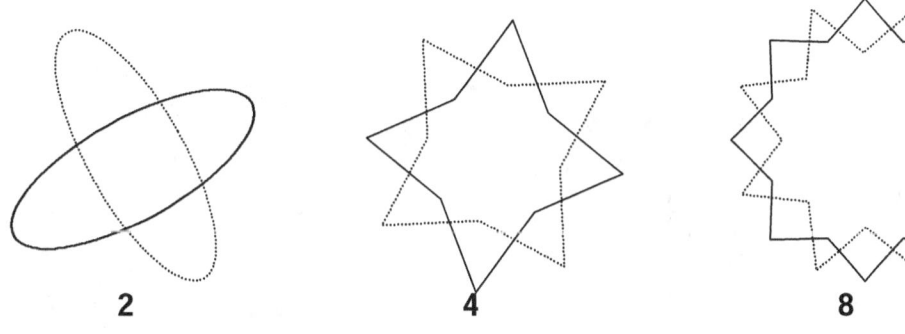

2   4   8

Notice that these resonant patterns are always a WHOLE number of peaks and troughs, which fit precisely between the two ends. The more excited the vibration pattern, the more energy there is; the more energy, the more mass from Einstein's $E=mc^2$. (Ref. 16.1, p.149)

Thus we see that a fundamental particle's mass and force can be measured by the exact resonance pattern of vibration of its internal string (Ref. 16.1, p.150) This contradicts the traditional idea that particles "contain" the "material" of its charge, mass, or force. An electron does NOT "contain" a negative electrical charge; rather, its string vibrates at a frequency that gives it its negative properties.

We now have an overall framework for the interplay of all the forces and matter of the Universe. Every "particle" of matter and every force are the result of the characteristic string whose pattern of vibration is unique to that particle.

Every physical occurrence or process in the Universe can now described by the harmony of vibrations of those strings representing the material of each of its constituents. (Ref. 16.1, p.156)

To achieve this extraordinary unification of matter and the forces of nature, scientists have proposed (and verified mathematically) that there are at least 10 more spatial dimensions, plus the dimension of time; but that is another topic for those wishing to delve more deeply into the modern advances in physics. (Ref. 16.1)

Did all of this intricate organization just fall into place by happenstance? . . . or is there a Supreme Intelligence that carefully created it?

Paul Garnett

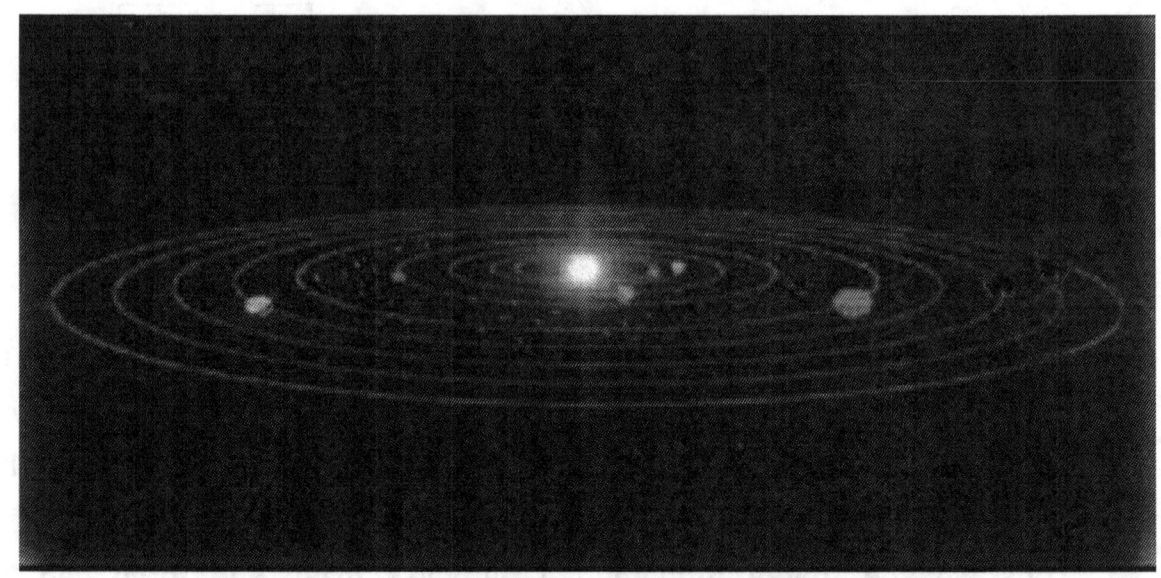

Order in the Solar System

Are the planets placed in a random order around the Sun?

Why is the Earth
unique
among all of the
planets in the
Solar System?

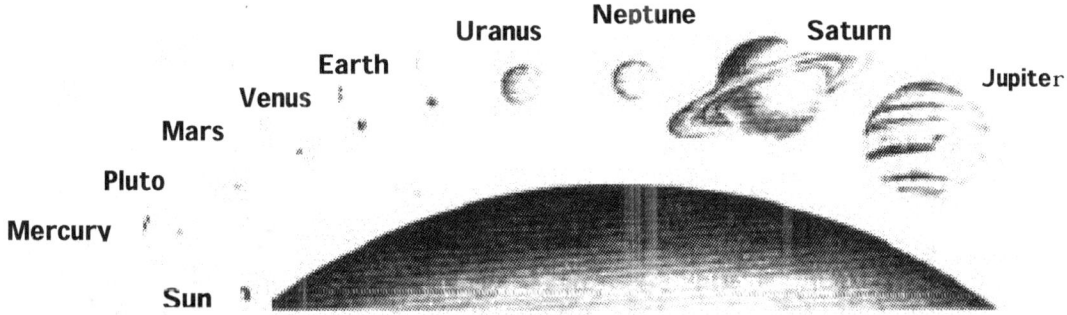

On a flat surface:

1) Shrink our **SUN** from 865,000 miles to 2 feet in diameter, represented by a *volleyball*.

2) At 164 feet away from the Sun, place a *grain of sand* to represent **MERCURY**.

3) Go another 120 feet away and place a *BB shot* to represent **VENUS**.

4) Mark off 156 feet more and place *a pea* to represent **EARTH**.

5) Go another 216 feet and place a *pinhead* to represent **MARS**.

6) Go another few feet and sprinkle some fine *dust* for the **ASTEROIDS**; then go 1576 feet and place an *orange* for **JUPITER**.

7) Go another 1868 feet and place a *golf ball* to represent **SATURN**.

8) Put on your hiking shoes and go another 4172 feet to place a *cherry* to represent **URANUS**.

9) After going 2 ½ miles, place *a dime* to represent **PLUTO**.

You now have a flat surface of about 5 miles in diameter. Representing our Solar System are: a *volley ball, a grain of sand, a BB shot, a pea, a pinhead, some dust, an orange, a golf ball, a cherry, and a dime.*

Using this same scale of measurement, you would have to travel another 6720 miles (distance across North America and back) to reach the nearest star.

> " *What is man that you are mindful of him? . . .Man is like a shadow,*
>
> *and his days are like a breath, lasting but a moment."* **Psalms 144:3**

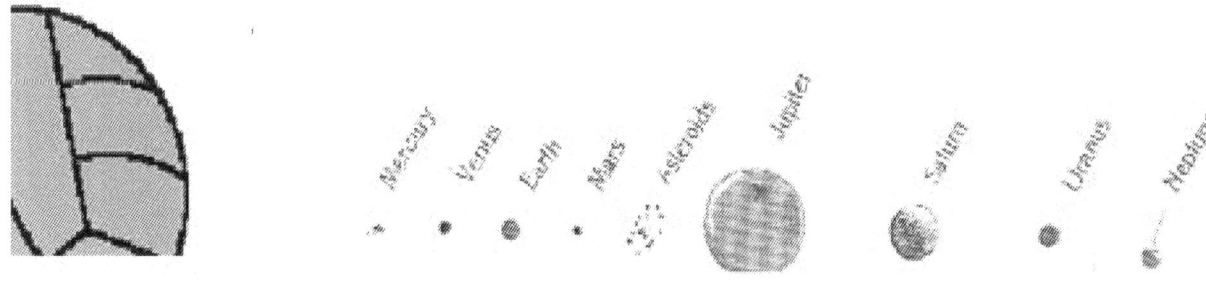

We have all looked up into the night sky; but with the latest satellite space telescopes such as the Hubble, we are able to see far beyond what the human eye can see through earth-based telescopes. These devices show us a universe that contains over one thousand Million galaxies, many contains over one thousand million galaxies just like our own Milky Way. Each of these galaxies contains between 1000 million and one million million stars similar to our Sun. This enormous complexity, so vast it is difficult to comprehend, none the less demonstrates an awesome unity, simplicity, AND ORDER held together by definite laws, principles and relationships

Mysterious formations in M-16.

What are these objects?

How did they form?

Where did they come from?

From the mega-structures of galaxies, which are millions of light years across to the atomic structure of DNA, which is too small to be observed and therefore must be deducted from X-ray shadows, all of these have the SAME basic structure. Did all of this happen by blind force and pure happenstance?

Much of what we have been considering will be new . . . NEW, not because you didn't make 'A' in your Science courses in school, but NEW because mankind has just recently been able to observe our environment through the new super-tech tools now available to see both farther out into Space and also objects infinitely smaller in size and finer in detail than ever before.

Earth from the surface of the Moon.

As you sit there in your seat today, you are traveling at about 1100 miles a minute, and that's about 1833 km/min, or 30 km/second. If the earth lost only 2 seconds out of every million miles it traveled, it would fly out of its orbit into the blackness of outer space. But it doesn't, and you don't, because God is at the controls . . . not chance or blind happenstance . . . using one of the Fundamental Forces (gravity) to keep you comfortably in your seat learning about His creation.

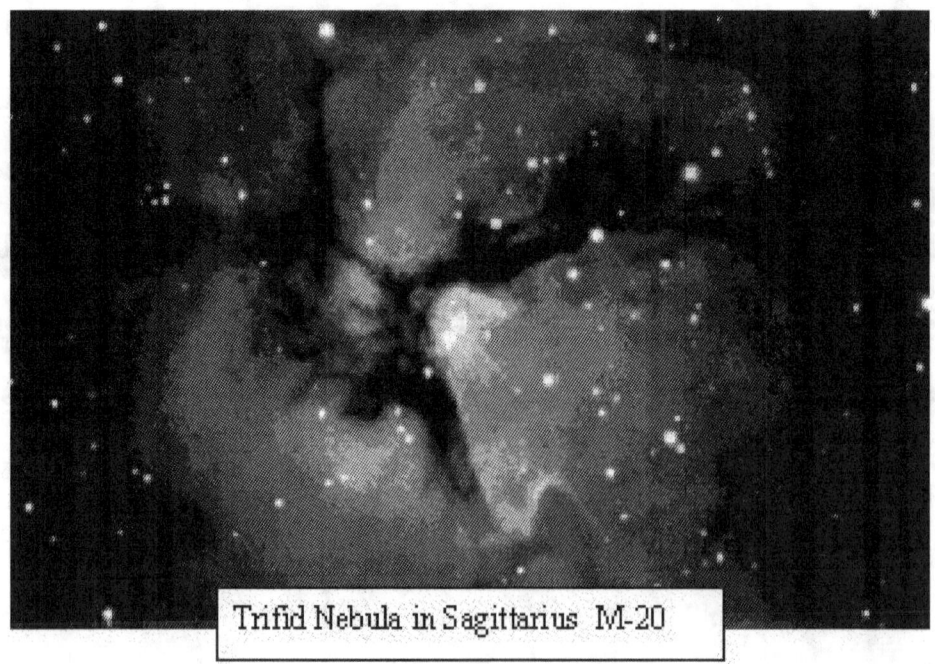

Trifid Nebula in Sagittarius M-20

Not only are the <u>Four Fundamental Forces</u> essentials here on Earth, but they also stoke the fiery furnace of our Sun. Let us consider **our SUN** . . . .

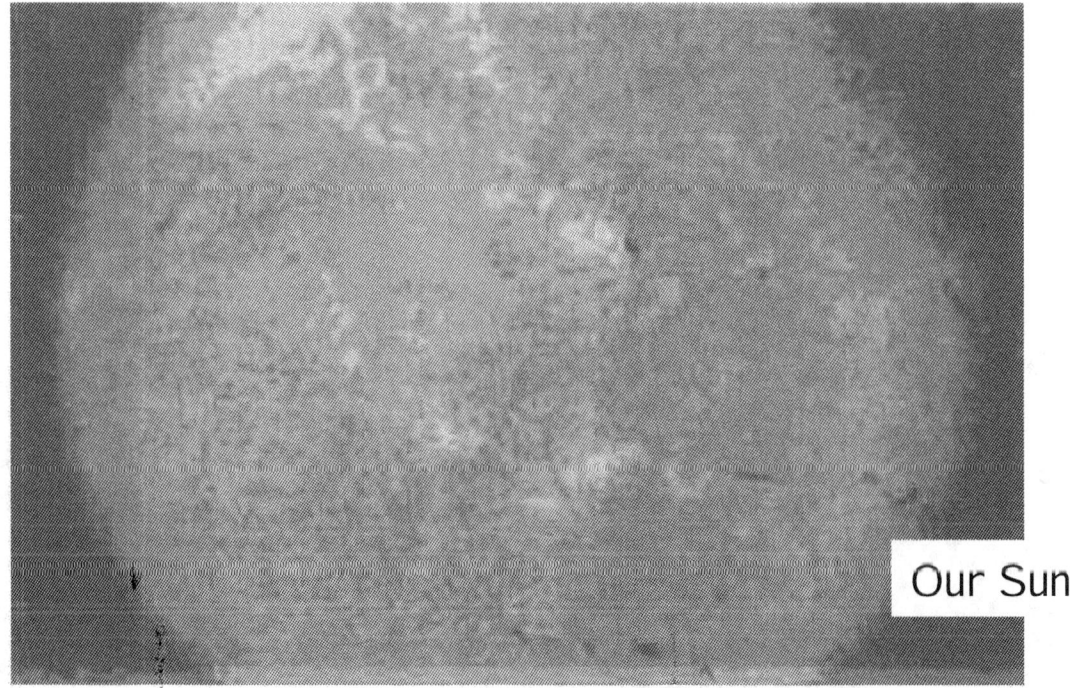

Our Sun

Our sun must provide stable, long-term radiant energy for life to exist on earth. This is accomplished by the fusion reaction, which we are familiar with, as the "Hydrogen Bomb" developed during World War II. When elements are fused together to form another element, there is also a tremendous release of energy; however, the exact values of the four Fundamental Forces of the Universe must control this series of precise fusion reactions in the formation of carbon and oxygen, necessary for all life forms.

Inside the Sun, under great pressure and heat, the precise amounts of quantum energy levels occur which permit these reactions to take place. These may be represented by:

1. Beryllium + Helium = Carbon

2. Carbon + Helium = Oxygen

Without carbon and oxygen life forms could not exist.

Arno Penzias, Nobel Prize Laureate, tells us: "The precise energy level in carbon is the result of the EXACT value of the Strong Nuclear Force and the Electromagnetic Force. There are more than 100 such cosmological balances between atomic and nuclear forces which make life possible on earth." (Ref. 21)

What is the planet closest to the sun? If you have difficulty in keeping the order of the planets straight, just remember the students' reminder sentence: **M**y **V**ery **E**xcellent **M**other **J**ust **S**erved **U**s **N**ine **P**izzas. From this sentence, you can see that the first planet is **Mercury**.

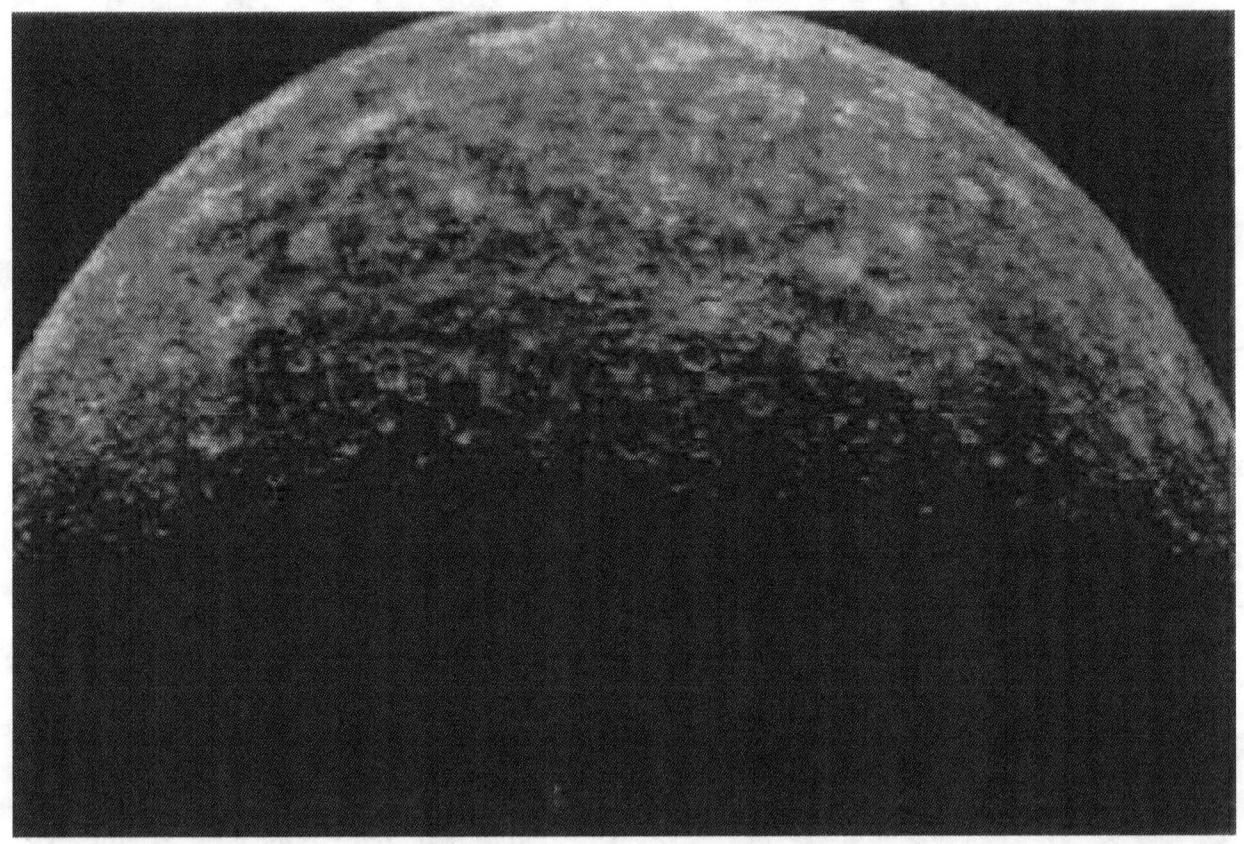

This sun-scorched little planet is closest to the Sun and looks very much like our moon, since both have no atmosphere, and they are scarred with the impacts of millions of meteorites. The contrasts in temperature from day to night are incredible: 415 C. in the daytime, which would roast an astronaut in his metal suit, and then plunging to –170 C. at night. Not a friendly place.

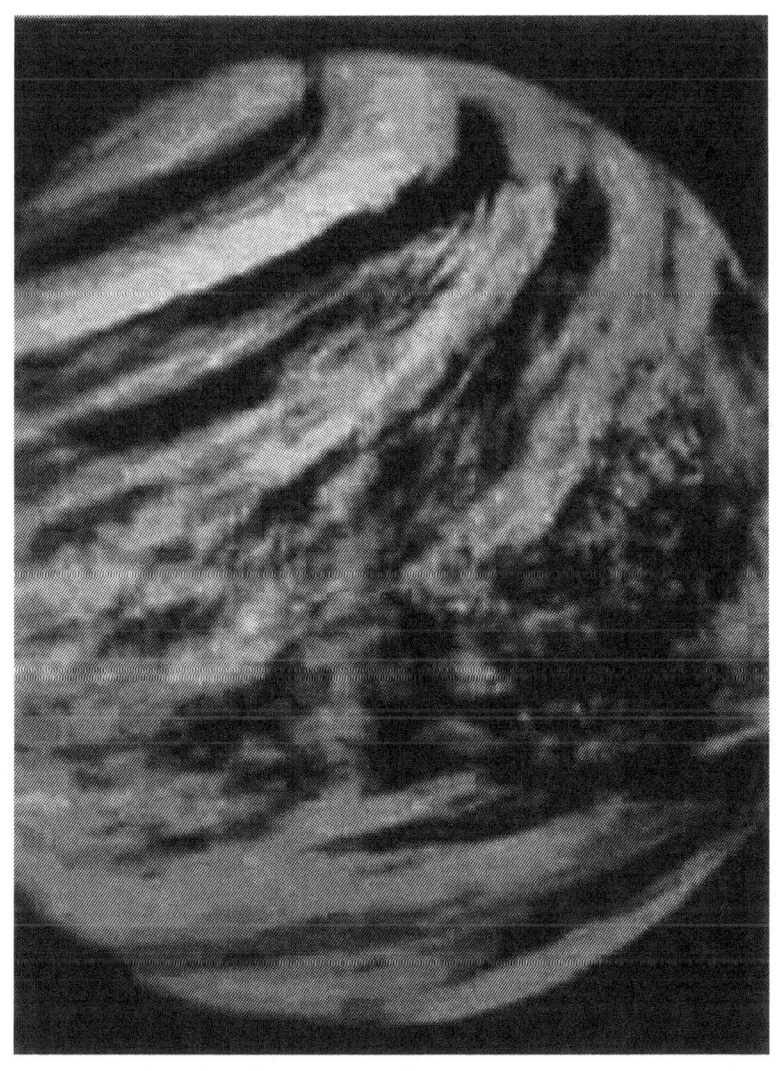

What is the next planet after Mercury?

It is Venus . . . . .

No one has seen the surface of Venus. Astronomers were long puzzled about the fuzzy appearance of Venus, until a Russian astronomer in the 18th Century showed that Venus has a heavy atmosphere of carbon dioxide. It wasn't until the mid-1950s that radio telescopes showed Venus to have a searingly hot surface, due to the Green House Effect. It is so hot that it will melt lead! And not only that, but in its CO_2 atmosphere, there are floating clouds of sulfuric acid. It is unlikely that any astronaut will ever step out onto the surface of Venus!

Do we need to ask what the next planet is?

Here is Earth, our beautiful blue-green home. Astronomers are universally agreed that the Earth is unique since it contains what no other planet in our solar system can possibly have . . . . . LIFE ! Both seas and land teem with life! On all of the other planets, the water has boiled away or is frozen solid. Only on Earth is there the delicate balance, which permits life to exist because of its orbit, its atmosphere, its iron core, its tectonic plates, and on and on, including many other factors. Probably the most important of these is the vast amounts of liquid water present on Earth.

This has happened because the distance of the Earth's orbit from the sun is exactly the correct distance, so that water neither boils nor freezes. The margin between boiling and freezing is very close, however, since Earth has dry waterless deserts and frozen ice covered poles. It wasn't always this way, as we will see as we look at the sequence of events when we examine the traditional view of Creation; however, the distance of the Earth's orbit is very critical.

Are the distances of the planets from the sun random? By chance? Apparently not! In 1772, a German astronomer, Johann Bodie, made a startling discovery regarding the distances of the planets from the sun. Here is what he found:

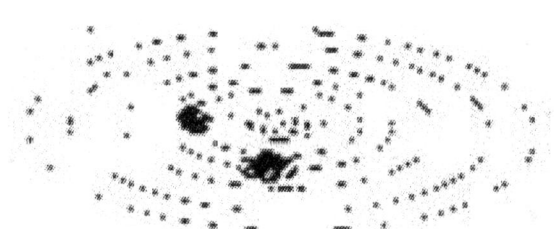

Bodie wondered if there was some definite order in the placement of the planets around the Sun. Several earlier astronomers had made proposals to that effect, but their telescopes were not sufficiently well developed to confirm their suspicions.

Like most people of his time, Bodie felt that the Earth must have a special place in the order of the planets. He also realized that the distances could not be measured in miles or meters or cubits or some other man-made measurement, but rather, some kind of astronomical comparative ratio.

For Bodie, that ratio must begin with the Earth because it was the most important planet. Astronomers knew the approximate distances of Mercury and Venus and Earth from the Sun, so Bodie began to experiment with what he knew.

He started by assigning the number 1 (or 10) to the Earth/Sun distance, and compared that number to the known distances of Mercury and Venus. After some experimentation, he discovered that if he started with .4 (or 4) for the distance of Mercury from the Sun, and kept that constant, then added the calculated distances of Mercury and Venus, he arrived at a table something like this:

Mercury   .4 + 0 = .4 (or 4 + 0 = 4)

Venus   .4 + .3 = .7 (or 4 + 3 = 7)

Earth   .4 + .6 = 1.0 (or 4 + 6 = 10)

Studying this table, he quickly noticed that, for each planet, the first distance number was doubling and that .4 (or 4) remained constant. If this was a fixed rule, he could predict the distance to the next planet. Prediction: .4 + 1.2 = 1.6 (or 4 + 12 = 16)

Bodie trained his telescope at the Earth/Sun distance of 1.6 (16), and there was Mars!

He had found a mathematical rule to place all of the planets in the Solar System!

Let's try it and see if it works.

MARS: What distance from the Sun would you predict for Mars?

4 + 12 = 16

We notice that Mars is not quite exactly the measured distance. There are good reasons for this, as we will discover in a moment. Mars is an arid red planet with a scant atmosphere of CO_2. Pictures taken from its surface by the Viking landers show a dead, cratered landscape. Mars is locked in an eternal ice age. Whatever water vapor may have erupted from its early volcanoes was instantly frozen, since its surface temperature remains far below freezing. Its orbit is far too distant from the Sun.

The USA is now planning a joint venture to the surface of Mars to follow the Viking Lander; however, to do this, the Space Station must first be completed and many technical problems must be solved.

At what distance would you predict the next planet?

4 + 24 = 28

When Bodie looked at this distance, he found NOTHING there. He thought there must be something wrong with his table of distances.

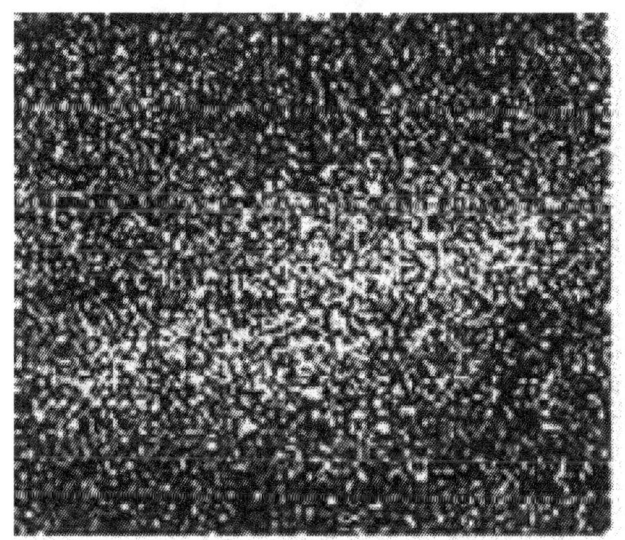

As we shall see, this mystery was later solved by another astronomer named William Herschel. Bodie, however, did the calculation for the next planet.

What distance would that be?

Jupiter: 4 + 48 = 52

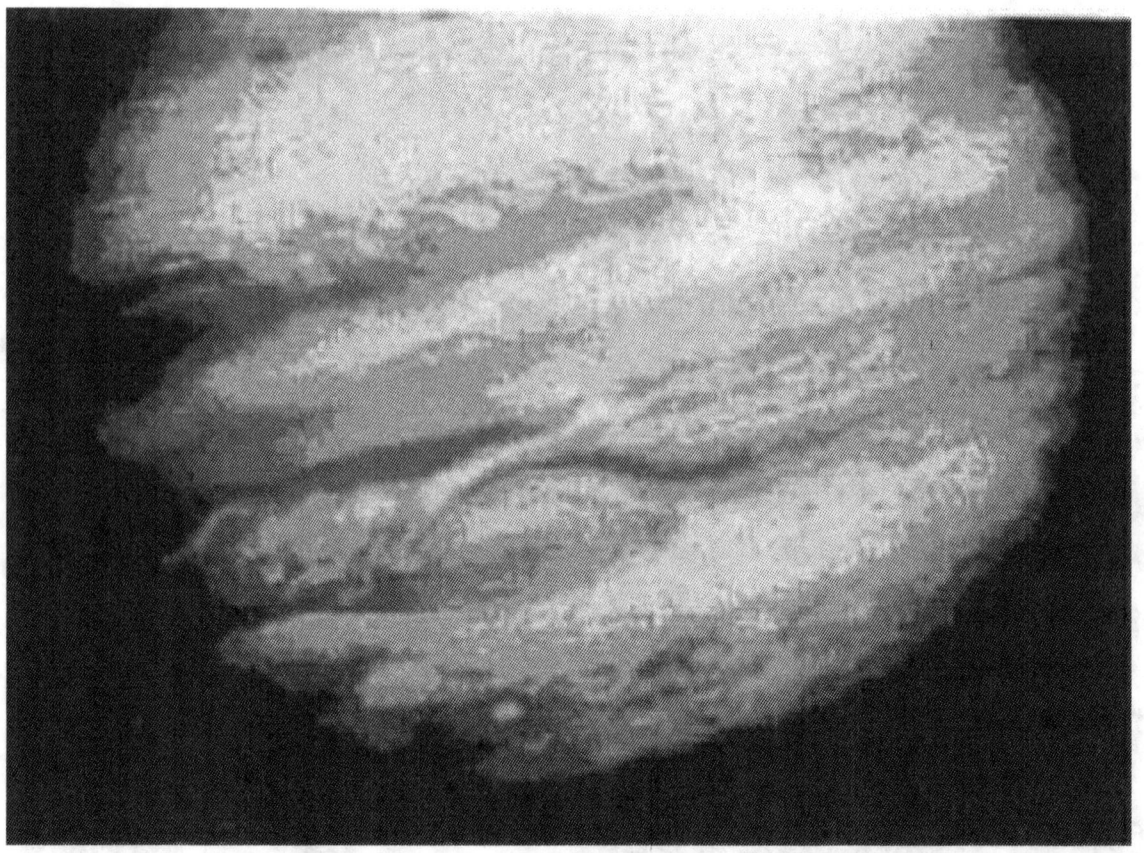

Jupiter belongs to the outer planetary group, being composed almost entirely of gas with a small rocky core. . .maybe. It is by far the largest planet with a size as wide as 11 Earths and outweighing all the other planets put together. The most prominent feature is the Great Red Spot, which is three times bigger than the Earth, and represents the top of huge hurricane that has been raging on Jupiter for at least the last 300 years.

The largest of its moons, Io, is coated with orange sulfur volcanic deposits, proving that there are volcanoes on its surface. Jupiter is considered to be a failed star. Too small to build up enough pressure to ignite nuclear fusion, and too big to form a solid rocky surface like the inner planets, it is called a "brown dwarf". It was made famous by the movie, *2001: A Space Odyssey*, where you may remember, the on-board computer, "HAL", takes over control of the space ship.

At what distance would you expect the next planet to appear?

SATURN: 96 + 4 = 100

Saturn is the lightest in weight of all the planets, and would float on water if there were an ocean large enough to hold it! Its most noticeable feature is the series of fine rings, composed of ice crystals and rocks.

Was Saturn the "Star of the East" that led the Wise Men to the baby Jesus? It is true that, in 7 BC, Saturn and Jupiter lined up; but at their closest conjunction, they were still about 2 moon widths apart, so it is unlikely Saturn could have been the Star of the Wise Men. Due to its size, however, it is the brightest planet in the sky and can be easily seen with the naked eye.

At what distance would you expect to find the next planet?

URANUS: 4 + 192 = 196

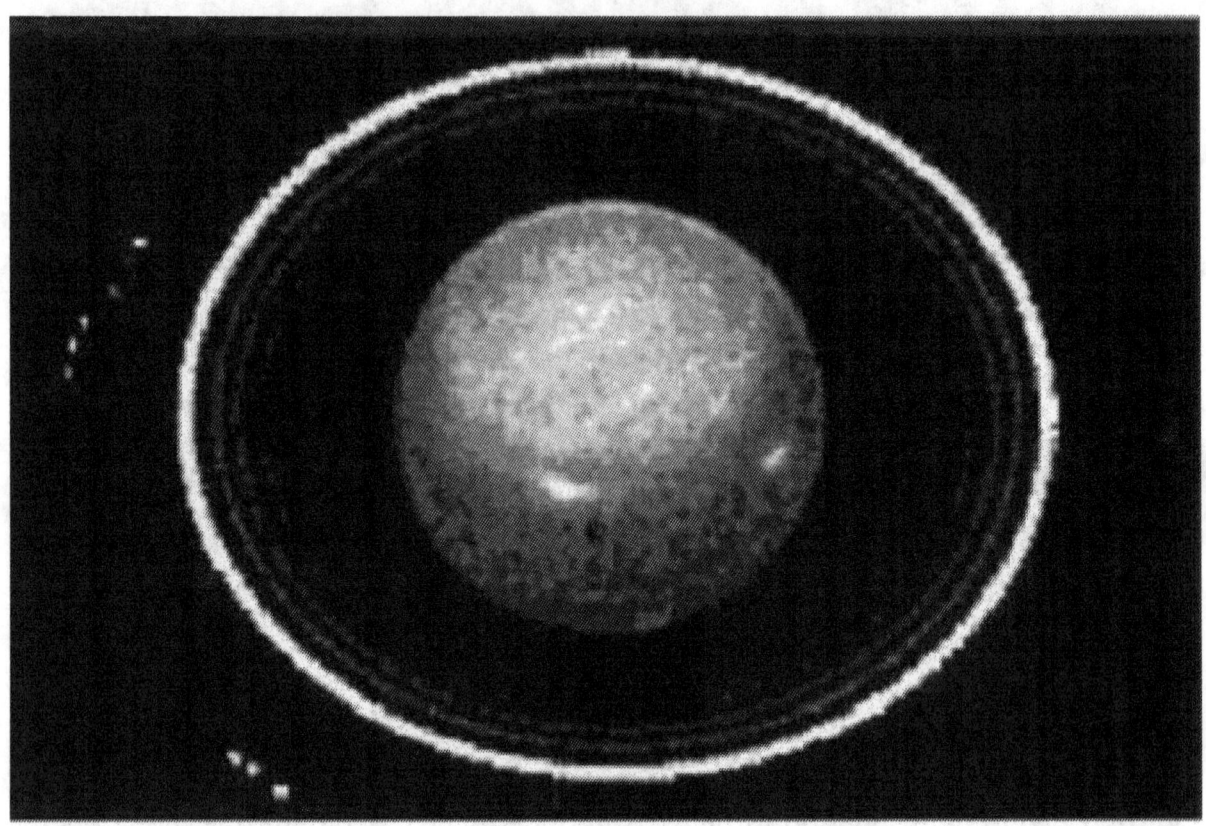

      Uranus is a curious, cold, distant planet, lying on its side. Its axis of rotation is towards us as we look at its pole, viewing it from Earth. Pluto and Uranus are considered by some astronomers to be "late captures" by our Sun since they do not match the plane of rotation of the other planets. Also, Uranus' axis of rotation is different than all of the inner planets.

**Question:
Whatever happened
to the missing planet
between Mars and
Jupiter?**

Having found Uranus with a much-improved telescope, William Herschel turned his attention to the blank space that Bodie had found in his table at the distance of 28 Earth/Sun lengths. The new telescopes revealed thousands of fragments that we now call the Asteroid Belt.

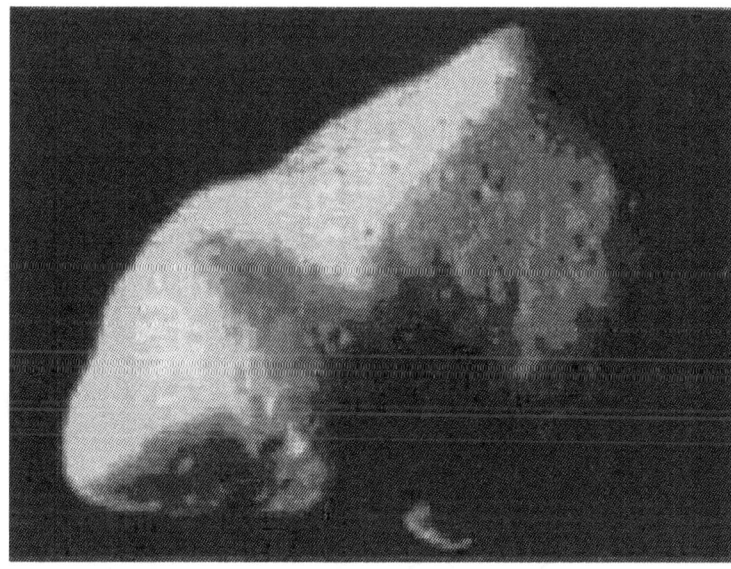

Asteroid Ceres

Asteroids arc the shattered remnants of a planet that may have exploded, or never quite formed in the first place. There are nearly 70,000 of them, although only about 4000 have been named and numbered. They are thought to be quite dangerous since many of them come close enough to collide with the Earth, although they could be redirected now if one is detected to be on a collision course with Earth. They are too small to be detected by the unaided human eye, but can be tracked by telescope as they approach the Earth.

No, the planets are not in some random, chance order around the Sun. They are definitely placed . . . . all except the outer planets.

Why is the distance from our Earth to the Sun the defining measurement of 10 or 1, which determines all the other numbers in Bodie's table of planetary distances?

Bodie's Numbers

Planet	Bodie's Numbers	Actual Distance
Mercury	4 + 0 = 4	3.9
Venus	4 + 3 = 7	7.2
Earth	4 + 6 = 10 (or 1)	10.2
Mars	4 + 12 = 16	15.2
(Asteroids)	4 + 24 = 28	28.3
Jupiter	4 + 48 = 52	52.0
Saturn	4 + 98 = 102	95.4
Uranus	4 + 192 = 196	192.0

Consider the thousands and thousands of possibilities for this distance of the Earth to circle the Sun.
What difference does it make? Why is it important?

It is important because it shows order in our solar system,

not random chance placement..

Looking into the Universe, we find order . . . order that is vast, beyond the scope of our comprehension, yet held together by fundamental forces that permit it to exist. This order allows life to multiply and inhabit the land, sea and air of this speck of dust in the endless spaces of the Cosmos.

And of all the life created on this planet, mankind has been chosen by the Creator himself to rule over all of the beasts both great and small, and to be touched by the hand of God himself.

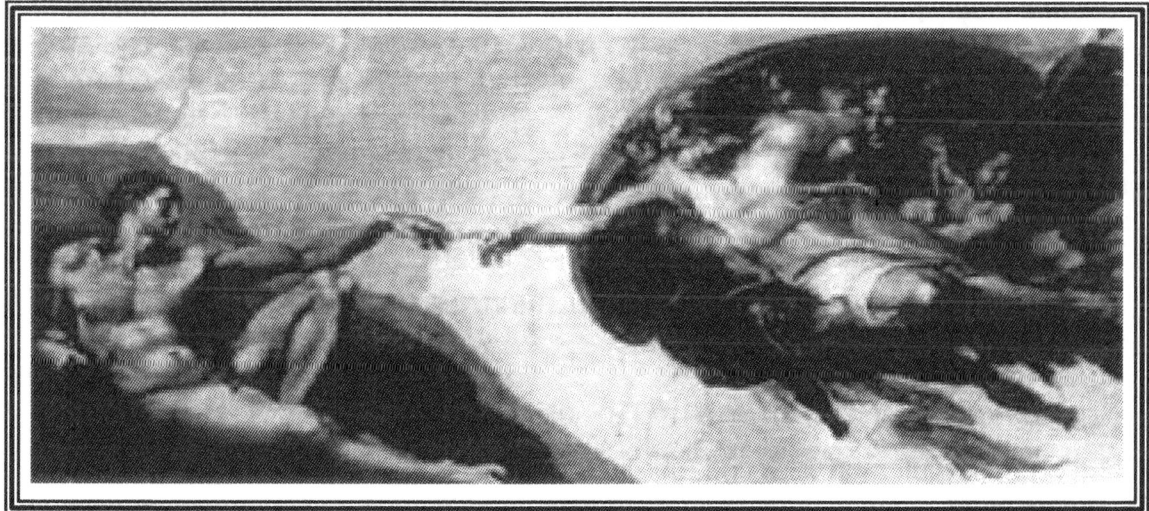

Another factor supporting Intelligent Design is the Cosmological Constant . . . .

Well, what is the Cosmological Constant? Quantum mechanics tells us that "empty" space is not really empty, but is filled with tiny particles of matter that are always suddenly appearing and then, equally suddenly, disappearing. They are called virtual particles and may be either positive or negative.

In any designated field of "empty" space, the Cosmological Constant tells us that the average energy value of the field is zero, average meaning that in some places the field may be positive, and some other places may be negative, but the average value is zero. Space is filled with virtual particles that blink in and out of existence, but the average value in any "empty" space is zero.

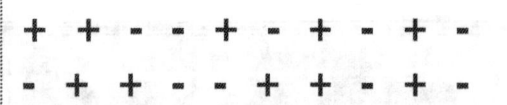

10 Positives

$$= 0$$

10 Negatives

Average charge is neither positive or negative, but is zero!

Why is it important that this Constant be zero? If it were some positive value, then this would produce an overwhelming of anti-gravity effect, preventing the formation of stars, planets and galaxies. If the Constant were negative, then this would produce a massive gravitational field, causing all matter to collapse in on itself; but instead, it averages zero, a fact that has amazed scientists, since any other value would instantly destroy our universe.

Cosmological Constant
Says: In any given measure of "empty" space the "average" energy is zero.

If it were a Positive Value

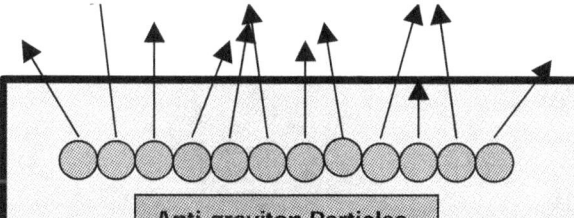

All forces of gravity cancelled by anti-gravity, preventing the formation of stars, planets, and galaxies.

An accumulation of anti-gravity particles would eventually destroy all "normal" particles in the Universe.

If it were a Negative Value

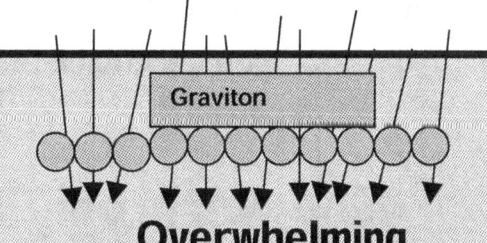

Overwhelming gravitational forces would cause all matter to collapse in upon itself.

An accumulation of high gravity particles would eventually cause a massive gravitational field, resulting in a universal black hole.

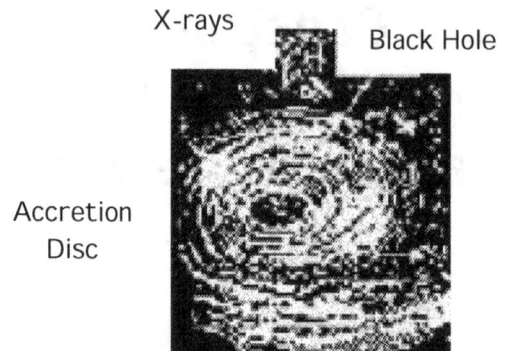

X-rays

Black Hole

Accretion
Disc

Because these forces are exactly the right amounts, our Universe has not been swallowed up by a gigantic Black Hole, nor has it collapsed into a fantastically hot nugget.

Instead, it is a perfectly balanced system, controlled by precisely explicit laws, which made possible the abundant life on Earth. Because of this fact, Nobel Prize Winner, Steven Weinberg, author of

the definitive text entitled, *The First Three Minutes* (Ref. 51) states the ANTHROPIC PRINCIPLE, a principle that says simply that we are here not by chance, but because the Universe is the result of coherent orderliness and the precision of the Four Fundamental Forces.

The Anthropic Principle states:

"If a universe does not take on the fundamental constants of physics which allow for the appearance and existence of intelligent life, then there never will be anyone to be aware of its properties; therefore it is reasonable to assume that the most probable universe is the one we are in, since it is so perfectly tuned to us."

This is a Nobel Laureat (physics) telling us that
the Universe is fine-tuned just for us here on Earth.

Why is the Earth Unique?

1) Its orbit is the precise distance for the sun to sustain life and liquid water.

2) Its speed around the Sun is exactly the proper acceleration of offset the gravitational pull of the Sun; less and it would fall into the Sun, more and it would fly off into outer space.

3) Earth's rotation permits the correct regular periods of light and dark to sustain plant (and thus, animal) life.

4) The Earth's atmosphere provides a protective shield against harmful radiation and the bombardment of meteors. It also acts as a reservoir of life-giving oxygen, nitrogen and carbon dioxide.

5) Our atmospheric blanket surrounds the Earth, keeping the Sun's warmth and protecting life from the chill of outer space (-270 degrees C).

6) This wonderful shield blanket, insulator is held precisely in place by a heavy molten core of iron, providing the exact amount of gravitational pull to prevent these light gasses from escaping into outer space.

7) The Earth contains vast supplies of water . . . the elixir of life. It is more abundant than any other substance in all its forms (liquid, solid, gas) because of the Earth's precise distance from the Sun. Water provides the basis of transportation of life's sustaining nutrients (in the form of blood and sap) for land life, and is the universal solution to support all life in the sea. This why all body fluids (i.e., blood and tears) have the same pH (alkalinity/acidity) as ocean water.

Water must possess very unusual properties to accomplish this array of requirements. For instance, as water cools, it becomes heavier and sinks, permitting the lighter warmer water to rise to the surface. Yet, as water approaches the freezing point, the process REVERSES! The colder water becomes ice and floats to the surface! The ice now acts like an insulator and keeps the deeper waters from freezing, thereby protecting the sea life beneath. Water is the only known compound that displays this characteristic. Without this unique quality, all the ice in the oceans would sink to the bottom of the sea, soon freezing all of the oceans into a solid mass.

All of this happened by chance?

Characteristics of the Universe that authenticate its design

All of these characteristics must be present for life to exist in the Universe.

Ross, Hugh 1998: Big Bang Model Refined by Fire. In *Mere Creation*, ed.William Dembski, 363-384. Downers Grove, Ill.: InterVarsity Press

1. **strong nuclear force constant**
    if larger: no hydrogen; nuclei essential for life would be unstable
    if smaller: no elements other than hydrogen

2. **weak nuclear force constant**
    if larger: too much hydrogen converted to helium in Big Bang, hence too much heavy element material made by star burning; no expulsion of heavy elements from stars
    if smaller: too little helium produced from big bang, hence too little heavy element material made by star burning; no expulsion of heavy elements from stars

3. **gravitational force constant**
    if larger: stars would be too hot and would burn up quickly and unevenly
    if smaller: stars would be so cool that nuclear fusion would not ignite, thus no heavy element production

4. **electromagnetic force constant**
    if larger: insufficient chemical bonding; elements more massive than boron would be unstable to fission
    if smaller: insufficient chemical bonding

5. **ratio of electromagnetic force constant to gravitational force constant**
    if larger: no stars less than 1.4 solar masses, hence short and uneven stellar burning
    if smaller: no stars more thin 0.8 solar masses, hence no heavy element production

6. **ratio of electron to proton mass**
    if larger: insufficient chemical bonding
    if smaller: insufficient chemical bonding

7. **ratio of number of protons to number of electrons**
    if larger: electromagnetism dominates gravity preventing galaxy, star and planet formation
    if smaller: electromagnetism dominates gravity preventing galaxy, star and planet formation

8. **expansion rate of the universe**
    if larger: no galaxy formation

if smaller: universe collapses prior to star formation

9. entropy level of the universe

> if larger: no star condensation within the protogalaxies
>
> if smaller: no protogalaxy formation

10.   mass density of the universe

> if larger: too much deuterium from big bang, hence stars burn too rapidly
>
> if smaller: insufficient helium from big bang, hence too few heavy elements forming

11.   velocity of light

> if larger: stars would be too luminous
>
> if smaller: stars would not be luminous enough

12.   age of the universe

> if older: no solar-type stars in a stable burning phase in the right part of the galaxy
>
> if younger: solar-type stars in a stable burning phase would not yet have formed

13.   initial uniformity of radiation

> if smoother: stars, star clusters and galaxies would not have formed
>
> if coarser: universe by now would be mostly black holes and empty space

14.   average distance between galaxies

> if larger: insufficient gas would be infused into our galaxy to sustain star formation for a long enough time
>
> if smaller: the sun's orbit would be too radically disturbed

15.   galaxy cluster type

> if too rich: galaxy collisions and mergers would disrupt solar orbit
>
> if too sparse: insufficient infusion of gas to sustain star formation for a long enough time

16.   average distance between stars

> if larger: heavy element density too thin for rocky planets to form
>
> if smaller: planetary orbits would become destabilized

17.   fine structure constant (a number used to describe the fine structure splitting of spectral lines)

> if larger: no stars more than 0.7 solar masses if smaller: no stars less than 1.8 solar masses
>
> if larger than 0.06: matter is unstable in large magnetic fields

18.   decay rate of the proton

> if greater: life would be exterminated by the release of radiation
>
> if smaller: insufficient matter in the universe for life

19. ^{12}C to ^{16}O nuclear energy level ratio

> if larger: insufficient oxygen

if smaller: insufficient carbon

20. **ground state energy level for ^{4}He**

　　if larger: insufficient carbon and oxygen

　　if smaller: insufficient carbon and oxygen

21. **decay rate of ^{8}Be**

　　if slower: heavy element fusion would generate catastrophic explosions in all the stars

　　if faster: no element production beyond beryllium and hence no life chemistry possible

22. **mass excess of the neutron over the proton**

　　if greater: neutron decay would leave too few neutrons to form the heavy elements essential for life

　　if smaller: proton decay would cause all stars to rapidly collapse into neutron stars or black holes

23. **initial excess of nucleons over antinucleons**

　　if greater: too much radiation for planets to form

　　if smaller: not enough matter for galaxies or stars to form

24. **polarity[7] of the water molecule**

　　if greater: heat of fusion and vaporization would be too great for life to exist

　　if smaller: heat of fusion and vaporization would be too small for life; liquid water would be too inferior of solvent for life chemistry to proceed; ice would not float, leading to a runaway freeze-up of all seasons.

25. **supernovae eruptions**

　　if too close: radiation would exterminate life on the planet

　　if' too far: not enough heavy element ashes for the formation of rocky planets

　　if too infrequent: not enough heavy element ashes for the formation of rocky planets

　　if too frequent: life on the planet would be exterminate

　　if too soon: not enough heavy element ashes for the formation of rocky planets

　　if too late: life on the planet would be exterminated by radiation

26. **white dwarf binaries**

　　if too few: insufficient fluorine produced for life chemistry to proceed

　　if too many: disruption of planetary orbits from stellar density; life on the planet would be exterminated

　　if too soon: not enough heavy elements made for efficient fluorine production

　　if too late: fluorine made too late for incorporation in protoplanet

27. **ratio of the mass of exotic matter to ordinary matter**

　　if smaller: galaxies would not form

　　if larger: universe would collapse before solar type stars can form

Our Infinite Universe

The diameter of our galaxy, the Milky Way, is such a vast distance that, if you could travel at the speed of light (7 1/2 times around the Earth in the time it takes to blink an eye), it would take you more than 100,000 YEARS to cross it! This is a distance so great it is difficult for us to comprehend it; yet our galaxy is just the BEGINNING of what astronomers can now see. Galaxies are said to be "as common as blades of grass in a vast meadow . . . at least 100 billion or more". ((Ref.14)

Discovery magazine (Oct, 1988) stated, "Astronomers and cosmologists are astonished by the coherent orderliness of the Universe, held together by the Four Fundamental Forces. . . . This is shown by the fact that, of all the human designed time pieces, none are as precise as the Universe."

All of this by chance?

Robert Millikan, discoverer of the electron, said, "The unbelieving astronomer must by mad!" (Ref.14)

"Lift up your eyes and look to the heavens. Who created all these? He who brings out the starry host one by one, because of his great power and might strength, not one of them is missing!"
        Isaiah 4:26

Fundamental Principles of Permutation, Probability and Chance
as applied to formation of DNA

Darwinian proposals vs. Intelligent Design

SEEDS

Do you like to travel? By sea? By air? Floating or flying like a butterfly? How about being shot out of a cannon like a circus performer? All these modes of transportation our Creator has integrated into the distribution of plant seeds.

Orchid seeds simply float off into the breeze like Tinker Bell's magic dust. Dandelions are equipped with parachutes. Maple seeds have wings that permit them to flutter like butterflies across the fields. Water plants provide their seeds with tiny air-filled floats; others are given sails, with tiny keels to tack to turn with the wind.

Perhaps the most amazing (and amusing) is a squirting cucumber that fires its seeds out from one end much like a cannon. The skin thickens while the inside continues to grow. The pressure inside increases until finally the stem is blown out like the cork of a bottle, firing the seeds out like a scatter gun.

Seeds also must be distributed at a precise time of the year, so they have an accurate calendar and moisture indicator to tell them not only the time of year, but also when the proper amount of water is available for their seeds to able to propagate. A desert plant not only waits for the precise day of the year, but also measures the amount of water available before releasing its seed package into a hostile environment.

All this by chance?

77

To better understand various recent advancements in science, we will have to look at some elementary concepts of Chance and Probability, plus the fundamental molecular biology of DNA and RNA. But don't panic, and stick with us. We can all do it together.

To help us understand the laws of Probability, let us begin with something we are all familiar with . . . a single die.

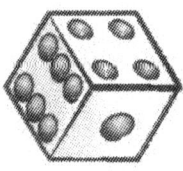

When we look closely at the die, we see that it has six faces, each one with a different number of dots on it from one to six.

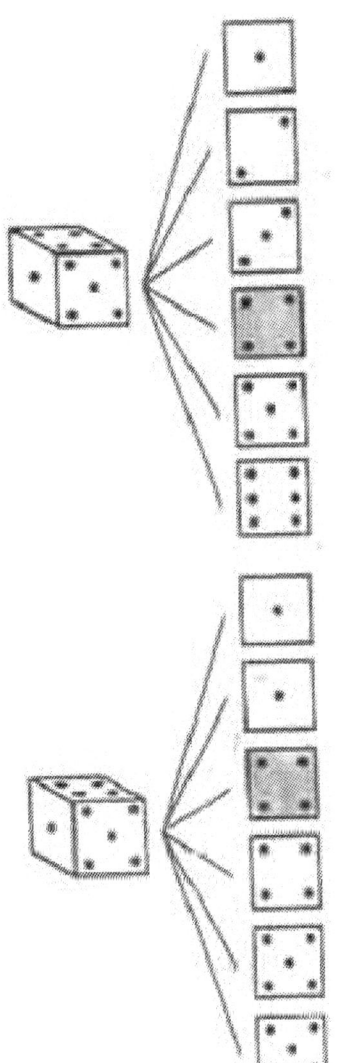

What are the chances of rolling a particular number, let's say, a 4? If we roll the die (electronically) a thousand times, we would discover the answer to our question is given by:

P (Probability) = The number of FAVORABLE ways divided by the TOTAL number of POSSIBLE ways.

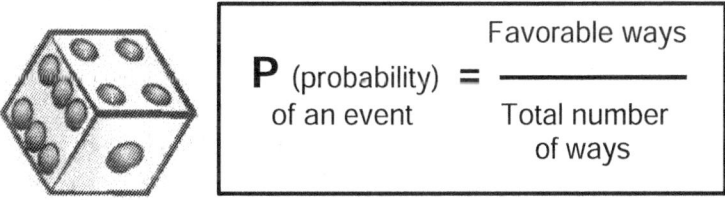

$$P \text{ (probability)} \text{ of an event} = \frac{\text{Favorable ways}}{\text{Total number of ways}}$$

In this case, P(4) = 1/6 or 17% chance of rolling a 4, since there is only one favorable way (the die has only 1 face with 4 dots on it), and the total number of possible ways is 6 because the die has 6 faces. So we have 1/6 – 17% chance of rolling a 4.

It immediately becomes evident that the chances are the same for each of the other numbers on the various faces of the die, since there is only one of each number on a single die, and the die has 6 faces.

What is the probability of rolling a 7?   P = 0 / 6, P = 0: There is no possibility because there is no 7 on any face of the die.

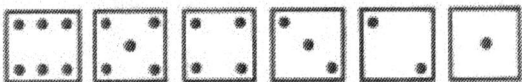

How about an even number?

P = 3 / 6 :   a 50% chance or rolling an even number because there are three even numbers and six sides : 3 / 6/= 1 / 2 = 50%

What about a number less than 10?

P = 6 / 6 P = 1 : A 100% chance, every time, because all of the numbers on all six sides are less than 10 : 6 / 6 = 1 = 100%

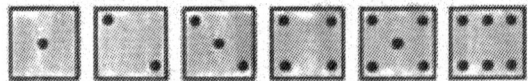

How about a Fibonacci Series Number?
(A Fibonnacci Number is the sum of the previous two.)
p = 4 / 6 P = 2 / 3 : A 67% chance of rolling a Fibonnaci Number because there are four Fibonnacci Numbers (1, 2, 3, 5) and six sides :
4 / 6 = 2 / 3 = 67%.

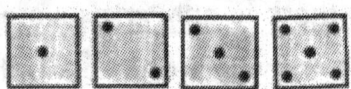

OK, this gives us some understanding of elementary probability; now let us see how these laws are applied. To understand their application, we must turn to the field of mathematics called

Permutations.

A Permutation is an arrangement of things in a DEFINITE, not random order.

Example: If there are two people seated at a table, then there are two arrangements in which they can sit: AB or BA.

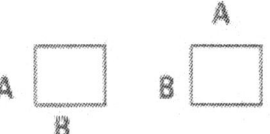

How about three people? How many ways can they be seated?

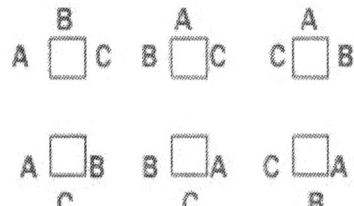

It turns out that there are six different ways for three people to be seated around a table.

When we do the same thing for four people, the number of ways jumps to twenty-four.

"There was a table set out under a tree in front of the house, March Hare and the Hatter were having tea at it: a Dormouse was between them, fast asleep, and the other two were using it as a pillow, resting their elbows on it, and talking over its head. . ."
(*Alice's Adventures in Wonderland* by Lewis Carol)

Is there some way that we can predict the number of specific ways that things can be arranged without making a list of all the different combinations we can think of, and then counting up the total?

Yes, this can be accomplished by what mathematicians call the Fundamental Counting Principle. To help us understand this, let us go back and consider the case of the four people sitting around the table.

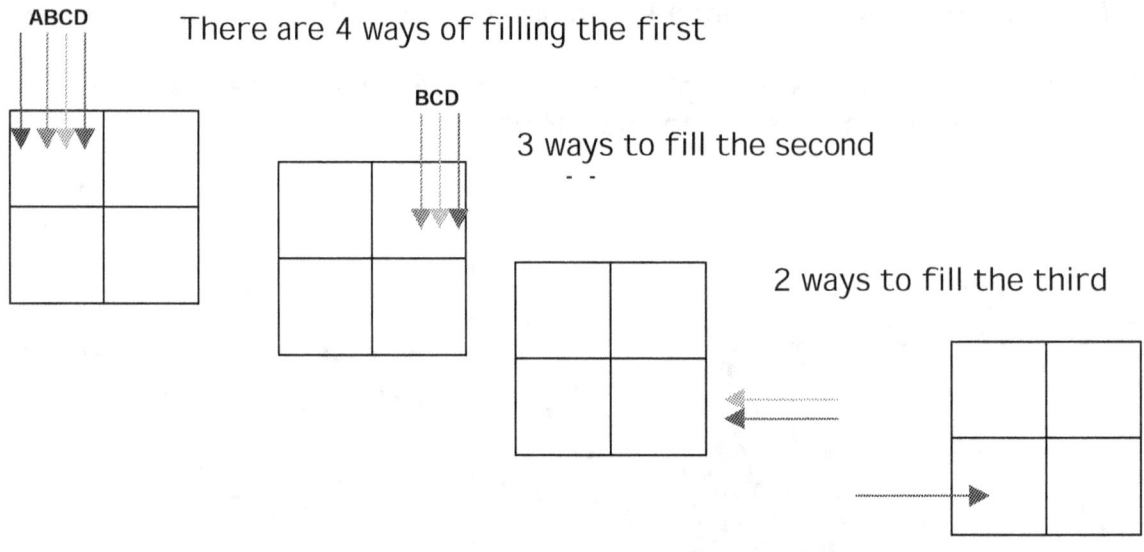

There are 4 ways of filling the first

3 ways to fill the second

2 ways to fill the third

Only 1 way to fill the last

There were four ways of filling the first position, since any one of the four people can sit there. Next, after the first position is filled, there are three ways of filling the second position (for the same reason), two ways of filling the third position, and only one way of seating the last position.

Each position at the table can be represented by boxes that look like this:

| 4 | 3 | 2 | 1 |

The number of arrangements can now be shown to be:

$$4 \times 3 \times 2 \times 1 = 24$$

Mathematicians represent this expression with a symbol that looks like an exclamation mark (!) which is called the factorial symbol, and is read: "Four factorial" which means: 4 x 3 x 2 x 1 . In general, for any number of different things, "n", the permutation of those things is: n! This means to simply multiply the consecutive numbers from "n" all the way down to 1.

$$4! = 24$$

How many ways is it possible for 14 people to sit on a single bench?

If you have a calculator, and you try to multiply it out, you may discover you won't have enough read-out space for the numbers. Although the factorial symbol is not meant by mathematicians to express surprise, you may be surprised at the answer. It turns out to be:

14! = 14 x 13 x 12 x 11 x 10 x 9 x 8 x 7 x 6 x 5 x 4 x 3 x 2 x 1

> . . .which equals 87,178,291,200 . . . more
> than 87 THOUSAND MILLION ways!!!!

In order for DNA helix to code for an organism, these nucleotides must be in the <u>exact correct order</u>, just as the letters in a word must be in proper order to convey the correct meaning.

Consider the word CREATION. If the position of just <u>one letter</u> is changed, then an entirely different word is formed: REACTION.

A single amino acid out of order in the DNA helix has the same effect of disabling the resulting protein. (Ref. 21)

For a single cell BACTERIA, imagine how many different ways 7 million people could sit around a table, keeping in mind the 87 billion different ways just 14 people could sit at the table.

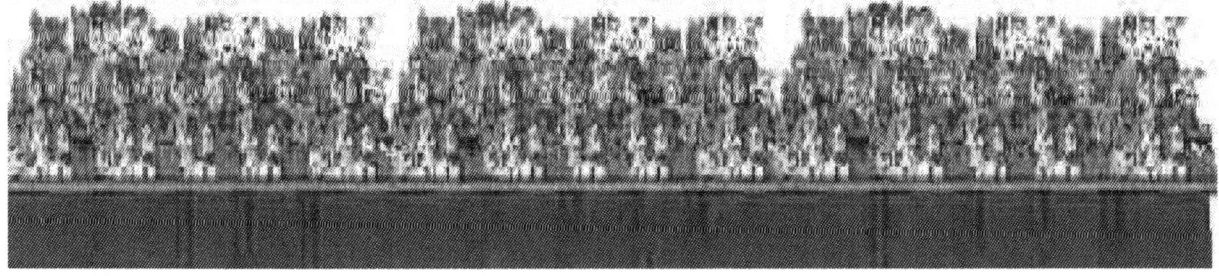

For just a tiny bacteria this number turns out to be a number so large that it cannot be calculated by most PCs.

This HUGE number now becomes the denominator in our simple formula for Chance

$$P = \frac{\text{Favorable ways (1)}}{\text{Total number of possible ways } (10^{65})}$$

The number 10^{65} is about the same value of the estimated number of atoms in our galaxy. (Ref. 21) This is for the chance occurrence for the correct sequence of nucleotides in the DNA for a single celled BACTERIA.

For a FUNGUS   we will need to calculate the permutation of <u>13 million</u> nucleotides in the correct order.

For a FLOWERING PLANT   we will need <u>several hundred million</u> nucleotides.

For a HOMINID we need about *3 billion* nucleotides.

As the numbers in the denominator grow larger and larger, the chance for any of these more complex organisms to occur diminishes, and then vanishes.

$$P = \frac{1}{1,000,000,000,000,000,000, \text{ etc.}}$$

You don't have to be a rocket scientist to see the mathematical model of the chance occurrence of a single "Simple" cell is impossible.   Yet this provable mathematical impossibility is being taught today to our high school biology students as scientific truth!

What happens when we put 1×10^{65} into our probability equation to calculate the chances of a "first" cell coming into existence just by random chance?
Remember P = The favorable ways it can occur divided by <u>the total number of ways it can occur</u>:

$$P = 1 / 4 \times 10 \text{ to the } 65^{th} \text{ power}$$

$$P = .25 \times 10 \text{ to the minus } 65^{th} \text{ power}$$

This is .25 <u>PRECEDED BY 65 ZEROS</u>. It is a number so small that it describes an amount that virtually does not exist.
It is a probability that is considered to be approaching "a zero order condition" by mathematicians. A zero order condition means statistically that an event is so unlikely to happen that it borders on the impossible.

000,000,000,000,000,000,000,000,000,000,000,
000,000,000,000,000,000,000,000,000,000,0025

Impossible? Most reasonable individuals would say that believing in something which would have the chance of 1 out of 4×10 to the 65th power of happening is pure wishful thinking, if not fantasy. Yet this fantastic tale is being taught in our schools today. .as fact.

Now, children, let me tell you a fairy tale about how all life on Earth began

Paul Garnett

The Molecular Organization of Life

and the mathematical impossibility of the Darwinian "first cell"

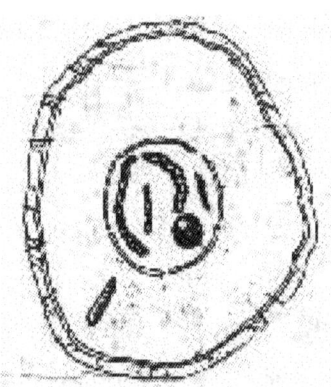

Miracle of Design in God's Universe

One of the smallest of all living things is a single celled organism called a diatom. Although microscopic in size, diatoms are in no way simple. They catch food, digest it, get rid of the wastes, move around, build houses, engage in sexual activity, and do all of this with no tissue, no organs, no hearts and no minds. Essentially, they really have everything we have and more.

They live in glass houses that they have built by absorbing silicon and oxygen from the sea water in which they live. In each tiny glass jewel box is a green bit of chlorophyll that not only sustains the diatom, but provides 90% of the food and oil for everything else living in the sea.

One of the thrills for biology students is to peer into a microscope for the first time to see these beautiful jewel boxes in a bewildering variety of shapes: circles, squares, ovals, shields, triangles, boxes, their glass surfaces always exquisitely etched with beautiful geometric designs, no two ever exactly the same.

All of this happened by the blind happenstance of evolution?
"That man is a fool who says to himself, "There is no God!". . . .(Psalm 14:1)

Evolution:

"Things are seldom what they seem,
  Skim milk often masquerades as cream"
W.S. Gilbert, *H.M.S.Pinafore*

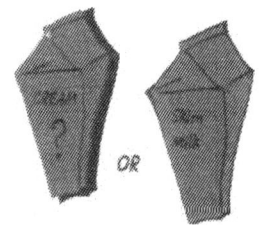

To sell the "Theory" of Evolution, its proponents have used a clever trick:
They combine the truth with a lie, a ploy often used by commercial advertisers
and dubious politicians.

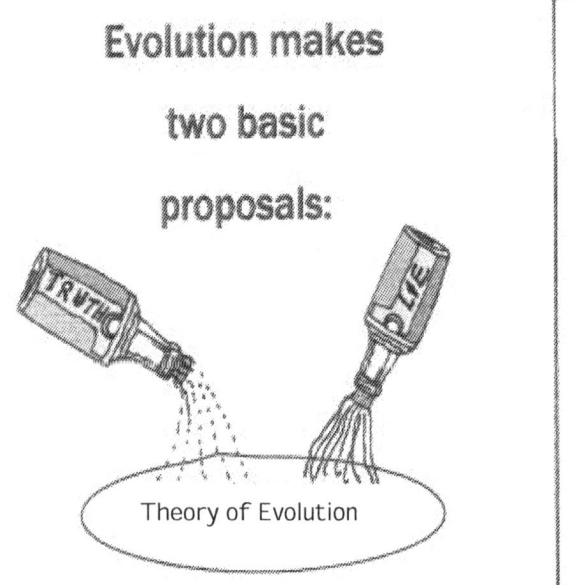

Evolution makes

two basic

proposals:

Theory of Evolution

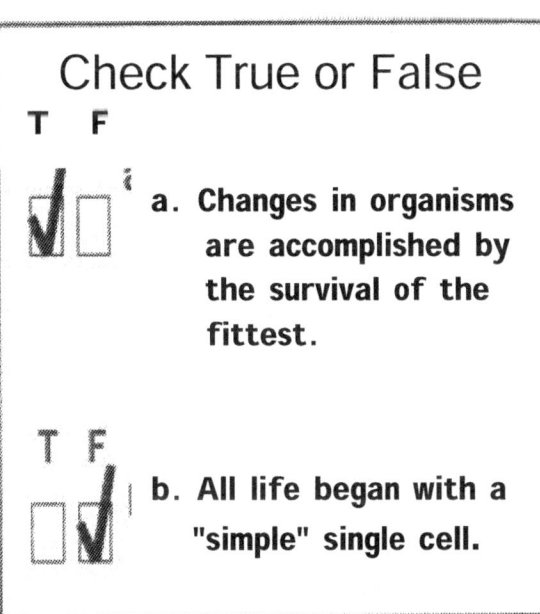

Check True or False

T F

a. **Changes in organisms are accomplished by the survival of the fittest.**

T F

b. **All life began with a "simple" single cell.**

The first of these propositions is essentially true and can be readily shown with a culture of bacteria in a Petrie dish and couple of drops of antibiotic. God, knowing that there would be essential changes in the environment, provided organisms with a mechanism to accommodate to those changes rather than to become extinct; thus we have "The Survival of the Fittest", a SLOW mode of adaptation to changes in the environment. This, however, is completely different from the proposal that all life began all by itself spontaneously in a single cell, and from this cell, all the millions of life forms "evolved".

86

According to Science, How did Life Begin?

Most scientists agree that the necessary pre-existing conditions for evolutionary chance origin of life must have been very complex indeed. It is assumed that the early atmosphere of Earth was quite hot in order to preserve the rare gases of krypton and xenon in the clouds. It is proposed that sunlight impinging upon the clouds, which contain water, carbon dioxide, ammonia and hydrogen, had a photosynthetic effect. This would have been much like chlorophyll does with present day plants: that of inducing the synthesis of amino acids, sugars and other molecules of greater complexity

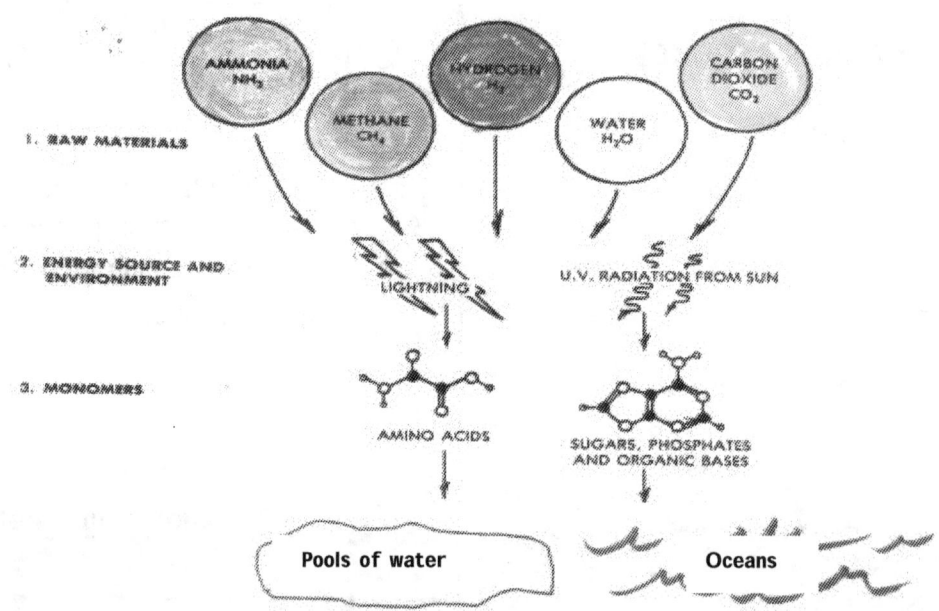

One year after this discovery by Stanley Miller in 1954, Nobel Laureate and Harvard University Professor of Biology, George Wald, made the following sweeping statement in the prestigious journal *Scientific American:*

"*However improbable we regard this event [the start of life], or any of the steps
which it involves, given enough time it will almost certainly happen at least once. And for life as we know it . . . once may be enough. (Ref.48.1)*

This statement appears in high school Biology books . . . <u>but what does NOT appear is the RETRACTION and APOLOGY</u> for its publication in 1979, the only ever made by any scientific journal of a Nobel Laureate's writing.

Consider the statement of world famous scientist Fred Hoyle from his book, *Astronomy and Cosmology* (W.H.Freeman, 1982, pp.540-541):

"In these concentrated pools, the building

of long chained molecules is thought to have

occurred, perhaps at the surfaces of particles of

clay. Over long periods of time, many combinations

are "tried, until at last the first self-replacing

system arose. This system became

the basis for all life."

"Many combinations were tried. . ." How many?

In order to answer this fundamental question we must first estimate how many "combinations" there must be for a "first" cell. We can do this by evaluationg what the absolutely necessary structures of a "first cell must be. David Deamer, biophysicist at the University of California (Santa Cruz) has already done this for us in his life-long search for a workable "first cell. Dr. Deamer's list includes the following:

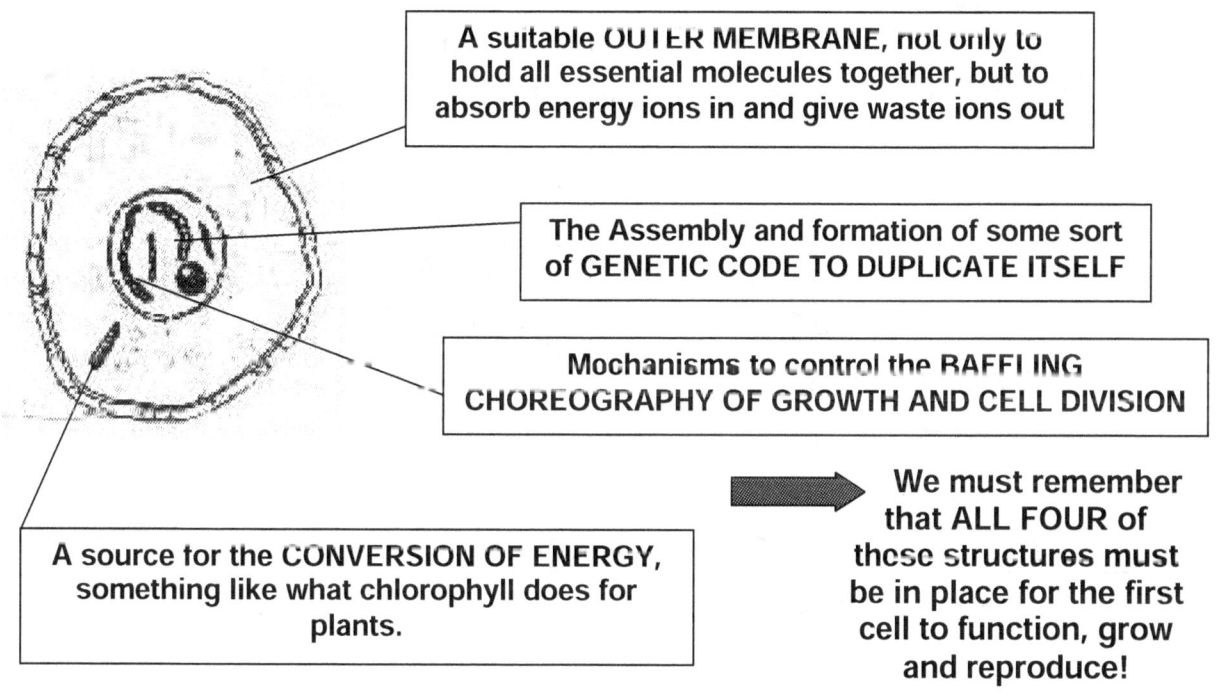

A suitable OUTER MEMBRANE, not only to hold all essential molecules together, but to absorb energy ions in and give waste ions out

The Assembly and formation of some sort of GENETIC CODE TO DUPLICATE ITSELF

Mochanisms to control the BAFFLING CHOREOGRAPHY OF GROWTH AND CELL DIVISION

A source for the CONVERSION OF ENERGY, something like what chlorophyll does for plants.

We must remember that ALL FOUR of these structures must be in place for the first cell to function, grow and reproduce!

88

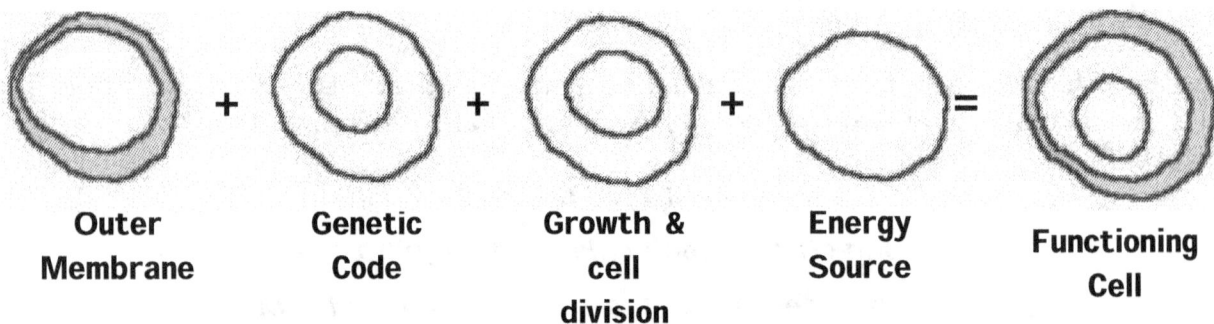

| Outer Membrane | + | Genetic Code | + | Growth & cell division | + | Energy Source | = | Functioning Cell |

We cannot begin with only a cell membrane, for instance, and then wait another million years for the next characteristic to develop, and so on for each of the four characteristics, to finally be present for the first cell to function. This is what is termed as *"irreducible complexity"*. Don't let this term frighten you. A system is irreducibly complex when it has several parts, which must closely work together for it to function. A good example of this is the ordinary mousetrap you can pick up at K-Mart. You will notice that there are several parts to this 89-cent item. First, there is a wooden slab acting as a base, a metal arm with a spring, a release catch in which to put the bait, and a metal bar, which restrains the arm.

In order to work, the system depends upon ALL OF THE COMPONENTS.

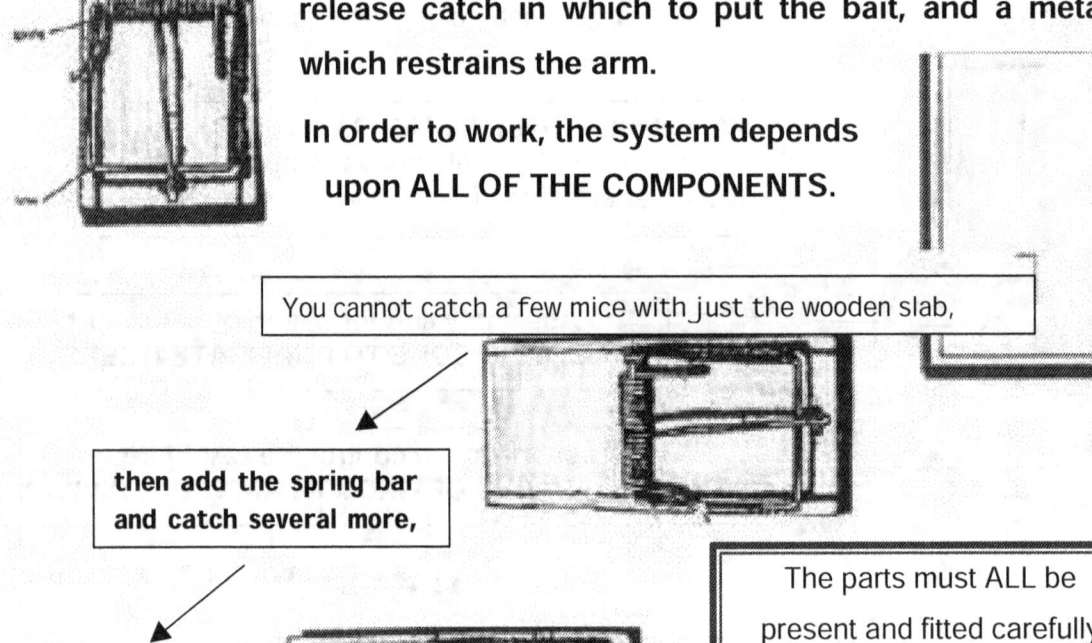

You cannot catch a few mice with just the wooden slab,

then add the spring bar and catch several more,

then add the bait release to catch even more.

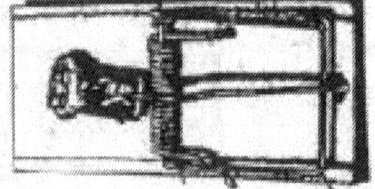

The parts must ALL be present and fitted carefully together to make the device work. That is what is termed "irreducible complexity.

What are the "irriducibly complex" systems of the first cell to make it function as a living thing? The building blocks, which compose cells, are proteins. How are proteins made? This is a question that baffled scientists for years, until Watson & Crick finally discovered the helical nature of the DNA molecule, which looks like a double spiral of alternating molecules of sugar and phosphorus, held together by a definite sequence3 of nucleic acids that look . . .

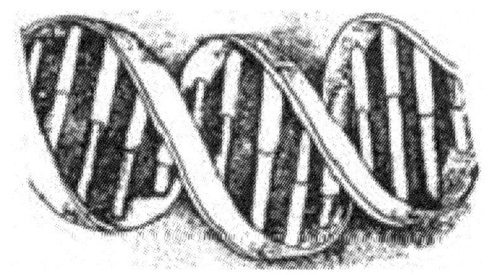

. . . something like this:

Does this look complicated?

Let us unravel the double spiral . . .

so that it looks like a
RAILROAD TRACK:

An easy way to view this structure is to
imagine a railroad track built of alternating
P & S molecules:
Imagine arailroad track railroad track built of
alternating P (phosphate and
S (sugar) molecules:

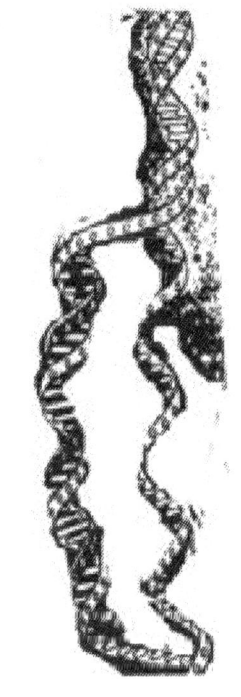

Sugar	Phosphate	Sugar	Phosphate	Sugar	Phosphate	Sugar

Sugar	Phosphate	Sugar	Phosphate	Sugar	Phosphate	Sugar

No railroad track is complete without the ties:

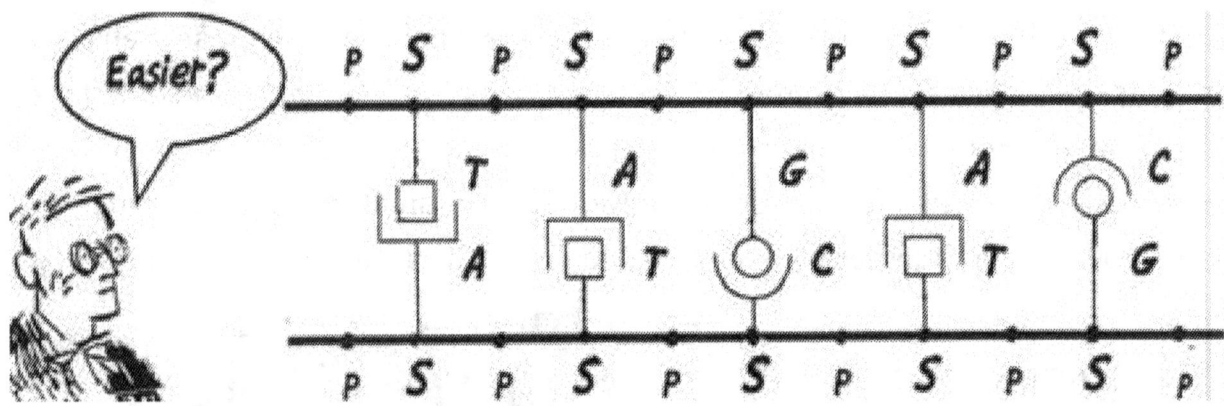

Now we add the ties to hold the tracks in place. These are made up of the four bases in double combinations as shown above. Notice that the ties join only the alternate sugar molecules and the base molecules **IN A DEFINITE SEQUENCE**, and that A matches T, and G matches C. We now have the chain of amino acids and the process of producing a life giving protein can begin. The steps for the protein forming sequence looks like this:

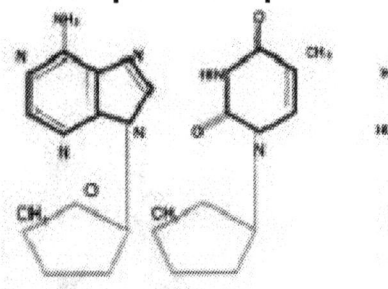
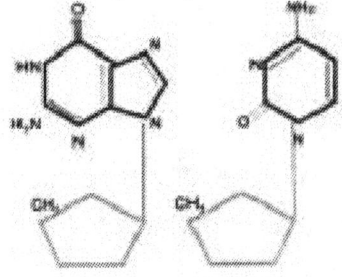

Adenine   Thymine   Guanine   Cytosine

The four Nucleic
Acids encode
for the
20 Amino Acids.

The Amino Acids

Alanine (Ala)
Arginine (Arg)
Asparagine (Asn)
Asparate (Asp)
Cysteine (Cys)
Glutamate (Glu)
Glycine (Gly)
Histidine (His)
I   soleucine (Ile)
L   eucine (Leu)
Lysine (Lys)
Methionine (Met)
Phenylalanine (Phe)
Proline (Pro)
Serine (Ser)
Threonine (Thr)
Tryptophan (Tyr)
Tyrosine (Tyr)
Valine (Val)

Which must be in

definite order in the

DNA

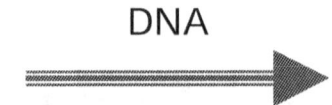

DNA   messengerRNA   PROTEIN made of amino acids

Transcription   translation

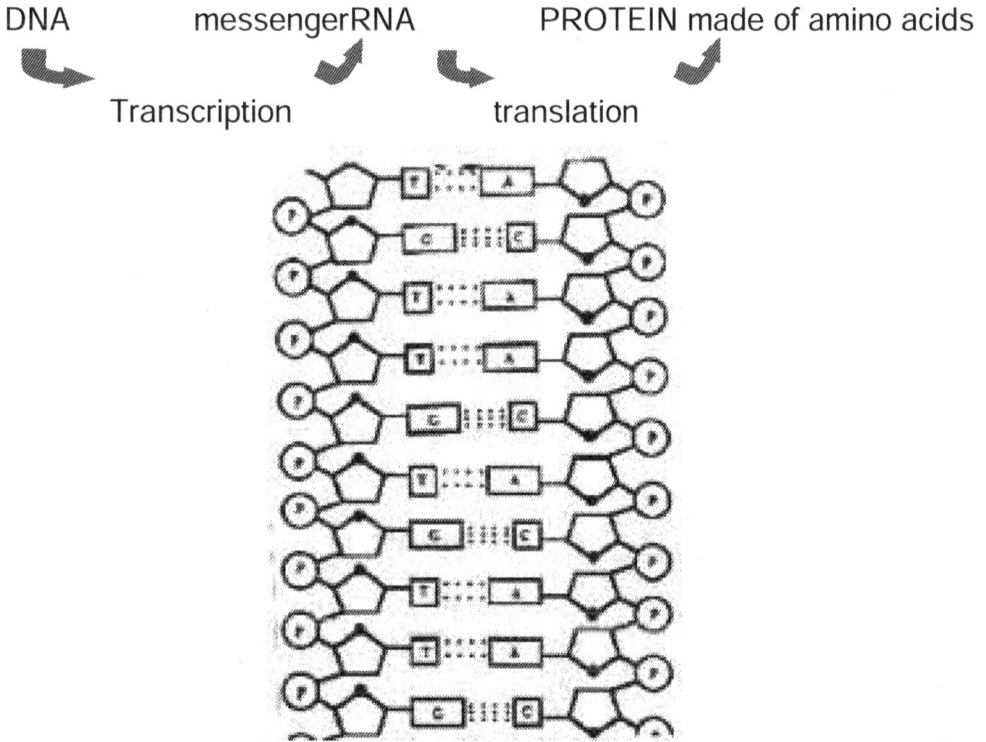

We remember that each protein in the cell is comprised of a long chain of definitely arranged nucleotides. (Ref.1) These nucleotides must be in exact order, just as letters must be in correct order to spell words, or words in a meaningful order to make a meaningful sentence.

Previously we have considered the word *CREATION*. If the position of just one letter is changed, an entirely different word is formed: *REACTION*.

Think now of the words in the sentence:

I *SAW a bear*.

Reverse the order of the letters in the second word, and the sentence becomes:

I *WAS* a bear.

A single nucleic acid out of order in the mRNA has the same effect of disabling the resulting protein. (Ref. 1)

| It IS possible to turn the arrangement upside down, but not possible (in order for it to work) to join C to T, or T to G, or A to G. If this happens, however, then the final result will be very different than if it remained right side up because only 1 strand is read, and then only from 5′ to 3′.

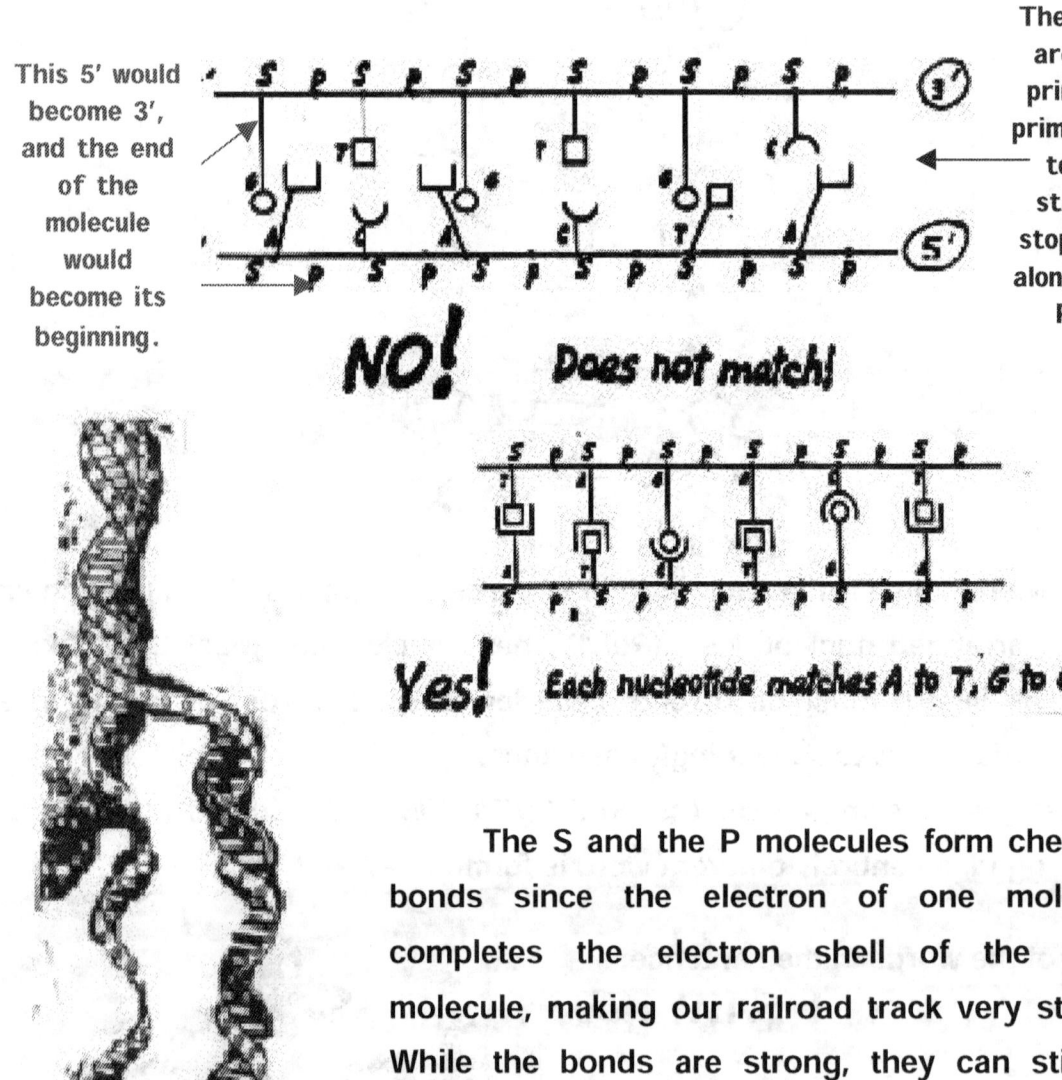

This 5′ would become 3′, and the end of the molecule would become its beginning.

These numbers are read: "5 prime" and "3 prime", and are to indicate starting and stopping places along the S and P tracks.

NO! Does not match!

Yes! Each nucleotide matches A to T, G to C.

The S and the P molecules form chemical bonds since the electron of one molecule completes the electron shell of the other molecule, making our railroad track very strong. While the bonds are strong, they can still be separated chemically.

The bonds between S and P molecules, however, are weaker than the bonds between the bases. When the connections of the bases are annulled one by one chemically and/or by heat, then the two halves come apart. When this occurs, we see an amazing thing happen.

When the two tracks of DNA and RNA are separated, the only way a new double track can be formed is to match EXACTLY the Sequence of the original DNA track.

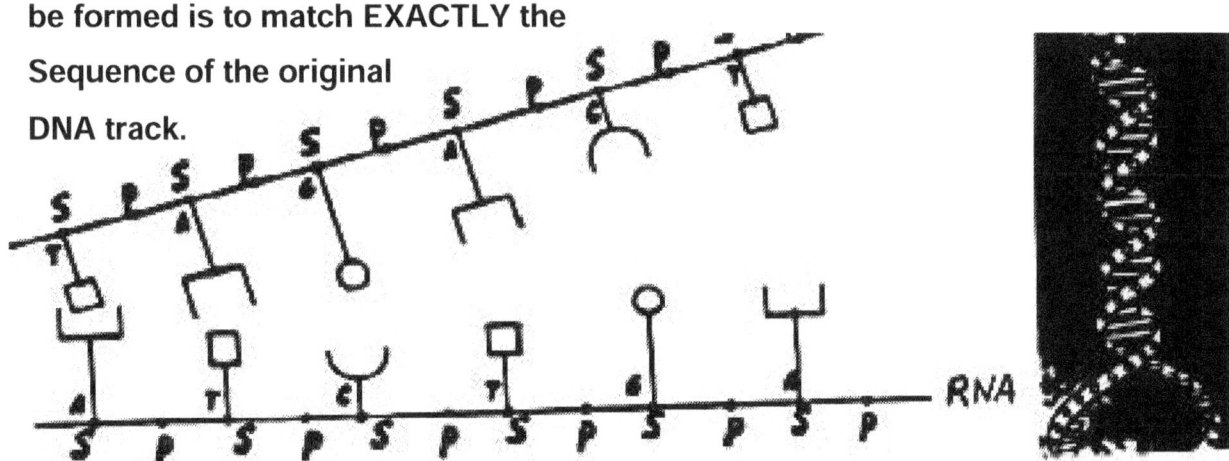

Assuming that there is an ample quantity of P and S molecules and of the four bases available, there is ONLY ONE WAY in which the second duplicate track can now be formed.

It must be reformed in an exact match of the original top track, since according to the laws of chemical combination, only A is permitted to combine with T and only G is permitted to combine with C. Our new double track will look like this:

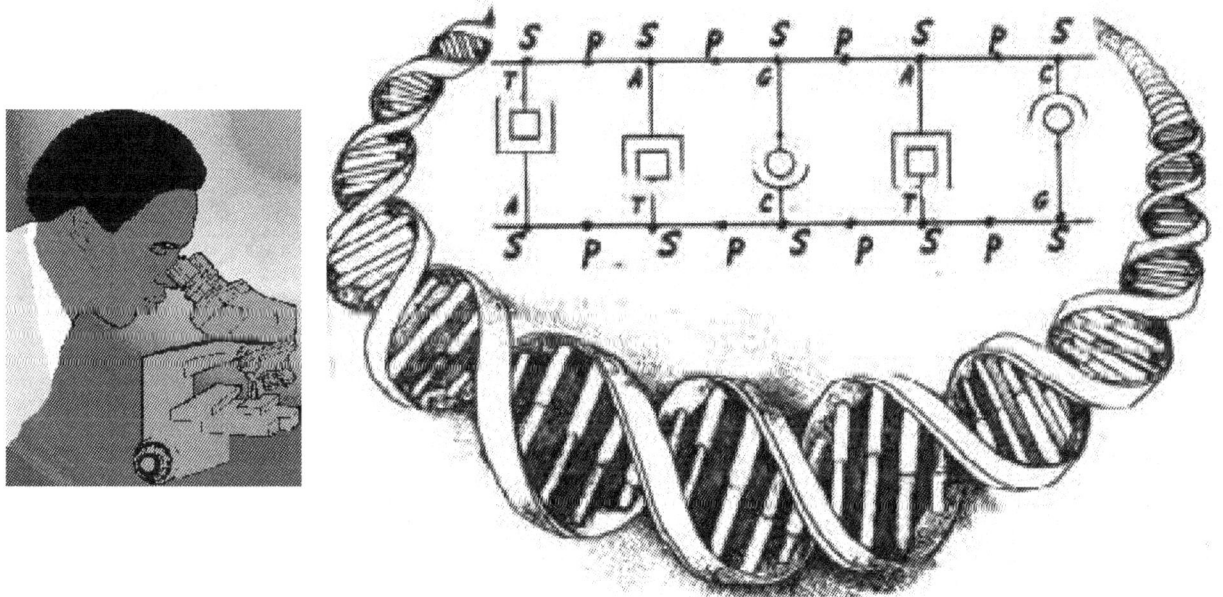

Not only does the double twisting helix spiral permit the efficient storage of genetic information, but its very design also permits the DNA to duplicate itself.

Let us now consider the construction of a typical protein molecule composed of 300 amino acids . . . *purely by chance.* No, instead let's go for just 100, and if we find that the formation of this much smaller number of amino acid chain (by chance) is impossible, then we will know that 300 is totally out of the question.

Let's look for a moment to see how the cell's building blocks are formed, and let us remember that there are four bases:

Adenine   Thymine   Guanine   Cytosine

There are four nucleotides (each of the four bases with a deoxyrobose or ribose sugar attached).

Adenosine   Thymidine   Guanisine   Cytidine

β-N-glycosidic bond

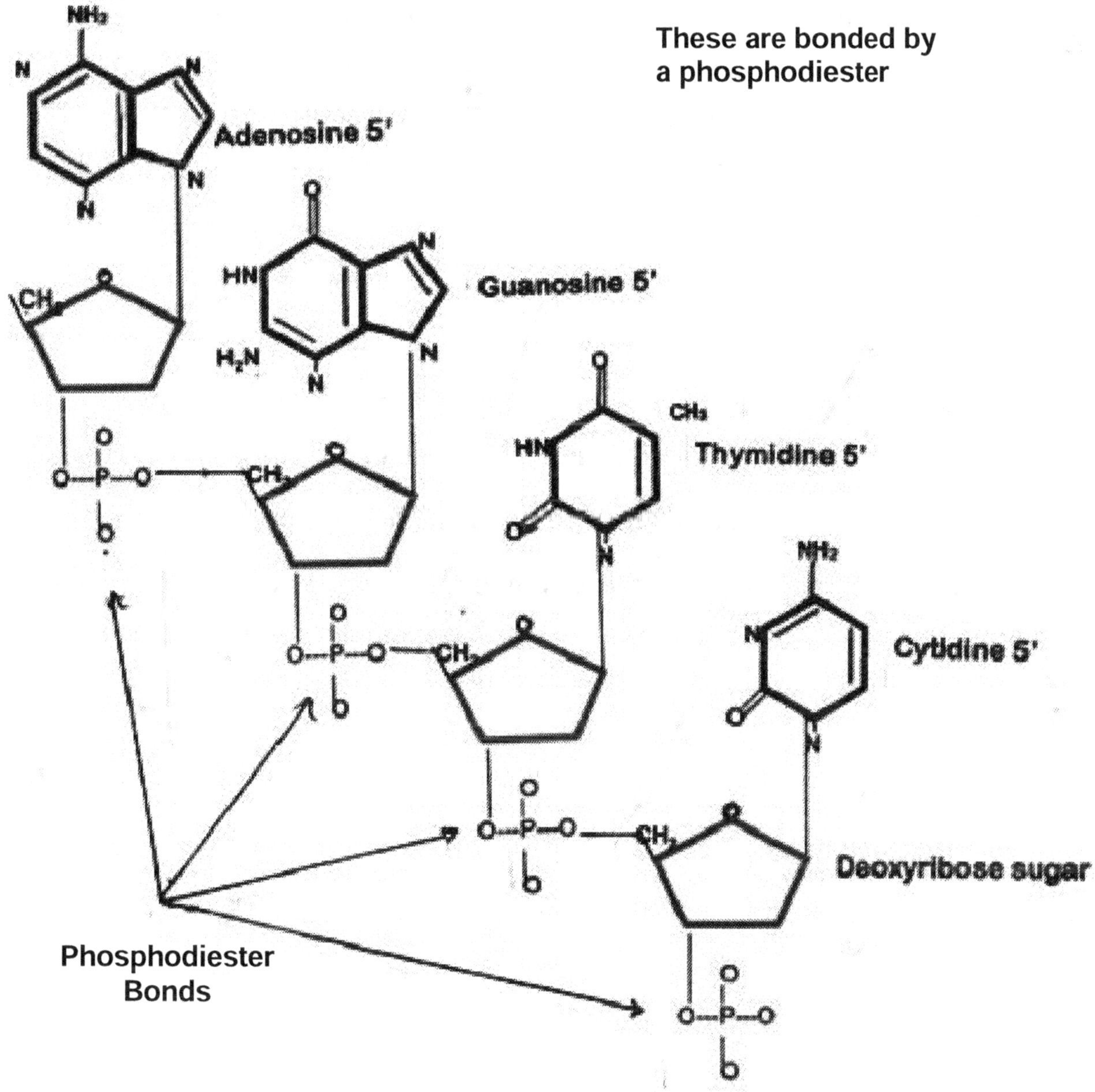

**These are bonded by
a phosphodiester**

Adenosine 5'

Guanosine 5'

Thymidine 5'

Cytidine 5'

Deoxyribose sugar

**Phosphodiester
Bonds**

For each amino acid, we need **3** nucleic acids to make up a protein. This nucleic acid trio is called a codon. There are **64** codons possible, making **64** factorial combinations, each one a different sequence. Of these **64** codons, **1** is needed to signal where the protein starts, and **2** are to signal where it ends; otherwise, the translation would never end. This would be like a sentence with no period or beginning capital letter.

Now, we must get the 100 amino acids to bond together in a chain by a chemical union known as a peptide bond. However, there are other bonds possible, and since only about one half of them are peptide bonds, the chance of successfully getting a peptide bond is about 50%. Thus, the probability of joining four peptide bonds is:

$$\tfrac{1}{2} \times \tfrac{1}{2} \times \tfrac{1}{2} \times \tfrac{1}{2} = 1/16 \text{ or } (1/4)\ 4^4 .$$

The probability of getting a chain of 100 amino acids, each with a peptide bond, would be $(\tfrac{1}{2})^{100}$, or about one chance in $(10)^{30}$.(Ref 28.1).

This is one chance in

1,000,000,000,000,000,000,000,000,000,000.

There isn't even a name for this number since there is no such thing as a zillion; however, to put it into perspective, the chances of winning the average state lottery are easily about 100 times better than this.

OK, let's say that despite the impossible odds, the new first cell DOES win the peptide lottery, and it now has a chain of 100 amino acids; however, the most difficult games of chance still lie ahead.

(L, left) **(D, right)**

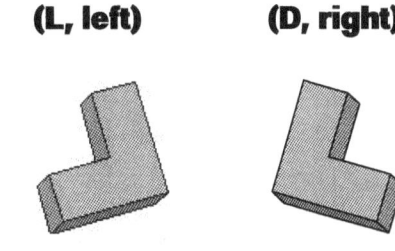

First we must consider that every amino acid has both a right hand version (D-form) and a mirror image left hand version (L-form). Only the left-hand versions work, although both occur in about equal numbers in nature.

To calculate the chances of getting all 100 L form amino acids in the chain, we must evaluate: (½) to the hundredth power, which turns out to be $(10)^{30}$ power.

So, the chances of building a 100 amino acid chain at random with all peptide bonds and all L-forms would be (¼)^{100} power, which comes out to be about one chance in $(10)^{60}$ power: one, followed by sixty zeros, or: 1,000,000,000,000,000,000,000,000,000,000,000,000,000,000,000,000,000,000,000,000. If we couldn't find a name for the previous number, there is little chance we will find a name for this one, so let's just call it a *ZAPTILLION*. Mathematicians call this a Zero Order possibility.

But we are not out of the chancy woods yet!

Remember the astounding number of permutations for just 14 people sitting around a table? We must now calculate the value of 20 factorial for each of the positions on the 100 amino acid chain! This turns out to be about 1 chance to $(10)^{130}$ power. There are only $(10)^{65}$ power atoms in our galaxy; we would need two galaxies of atoms to get an idea of the size of this number.

Biochemist Michael Behe has compared the possibility of forming a 100 amino acid chain protein to be about the same as that of a blinded folded person wandering through the vast Sahara Desert to find one single marked grain of sand, and having to find it not once, but three times!

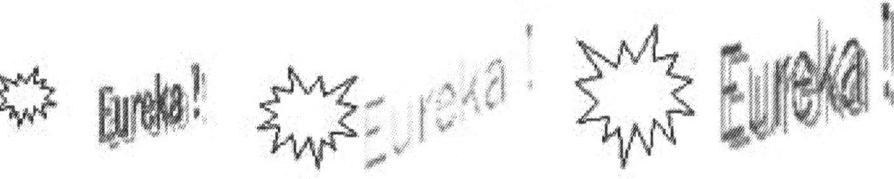

When the amino acid chain is in the correct position, the chemical interactions between the amino acids cause the chain to take on a 3-dimensional shape.

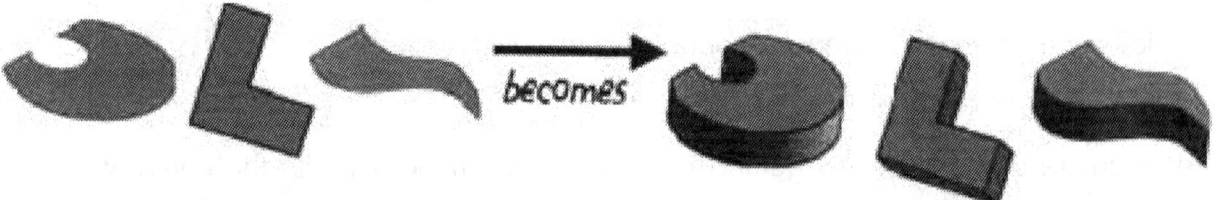

becomes

Shape of amino acid chain

3-dimensional because of the electro-chemical interaction of the molecules

This shape determines what function the amino acid chain may have in the cell and how it may fit with other amino acids to allow the DNA to coil itself into the double helix. The fitting arrangement of the amino acid chains is part of the information storage of the DNA design.

Computer chip . . .

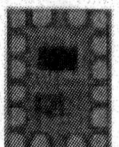

The miracle of our age

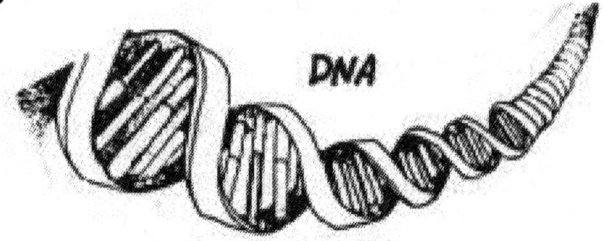

DNA

The miracle of ALL ages

Living DNA is several trillion times more efficient than our present most advanced computer chips when it comes to storing information (Ref. 15.2)

C++ Computer Code

```
// input stream contains
// "The quick brown fox! - jumped or not"

char buffer[8];
cin.getline(buffer, 10, ' ');
// get: "The quick brown fox" into buffer
char c;
cin.get(c);
// get: "!" is c
cin.get(); cin.get(); cin.get();
char buf[6];
cin.get(buf, 6);        // gets "jumped" into but
int i = cin.gcount(); // i gets 6
```

Hebrew Biblical Text

‏1. סוֹסוֹת הַמֶּלֶךְ הַטּוֹבֵהַ 2. לָקַח הַנָּבִיא סוּס אֲשֶׁר לַמֶּלֶךְ |
‏3. אֵלֶּה יְמֵי אֲשֶׁר מֵת הַמֶּלֶךְ קָרֵב 4. נָתַן אֱלֹהִים אֶת־הָאָדָם
‏5. אָמַר עָבַד אַבְרָהָם אֹכִי 6. וַתֵּל־פְּנֵי כָּל־הָאֲדָמָה הָעַמִּים
‏7. אָמַר אֶת־לְבַב הַמֶּלֶךְ הַגָּדוֹל מְרַע הַנְּבִיאִים יָשֵׁן
‏8. לֹא שָׁמַד קָם אֶת־דְּבָרֵי וְנִשֵּׁי הָעֲלִיסִים 9. חָרַב הֶחָכָם בְּיָד
‏10. לֹא וְקָרְאָם אֵת מִצְרַיִם אֲשֶׁר אָמַרְתִּ הִי נְבִיאֵי הָעוֹבֵד
‏הָעֲלִיסִים

Information contained in either a written language or in computer codes is the result of Intelligent Design, not random chance.

99

Given the original track, containing all the correct sequence of bases, the living cell can now duplicate itself by utilizing the PRECISE SEQUENCE of the four bases A, T, G, and C in the double helix of DNA/RNA.   Beware! Should the original sequence be destroyed or confused, there is no way the plant or animal cell can somehow reassemble it together again in the proper sequence since the numbers required to get the correct sequences are so great.

Think of trying to reassemble six thousand million combinations of the human genome in the EXACT CORRECT ORDER, even if we knew what the correct order is, which we don't quite. . . yet.

The amount of DNA in a cell varies roughly with the complexity of the organism.   Bacteria have about several million nucleotides of DNA. The amount of eukaryotic DNA ranges from a low of several tens of millions of nucleotides in fungi to a high of several hundred million in some flowering plants. Humans come in at around three billion nucleotides.

The links within all pairs of bases (A & T) and (G & C) can be broken by radiation or disease. Breaking of the bonds, called mutations, can cause mismatching of the four bases. Most insertions into the complex sequence of genetic information usually tend to garble the content rather than improve it. This is not always true, however, because in some unusual cases, the changed plant or animal is better able to cope with its environment, and in these cases, the organism is improved. Nevertheless, most mutations are lethal or harmful, and to maintain that this random process is the mechanism, by which the complex sequence of genetic information was arranged in the first place, is to ignore the Laws of Probability and Permutations, as we have shown.

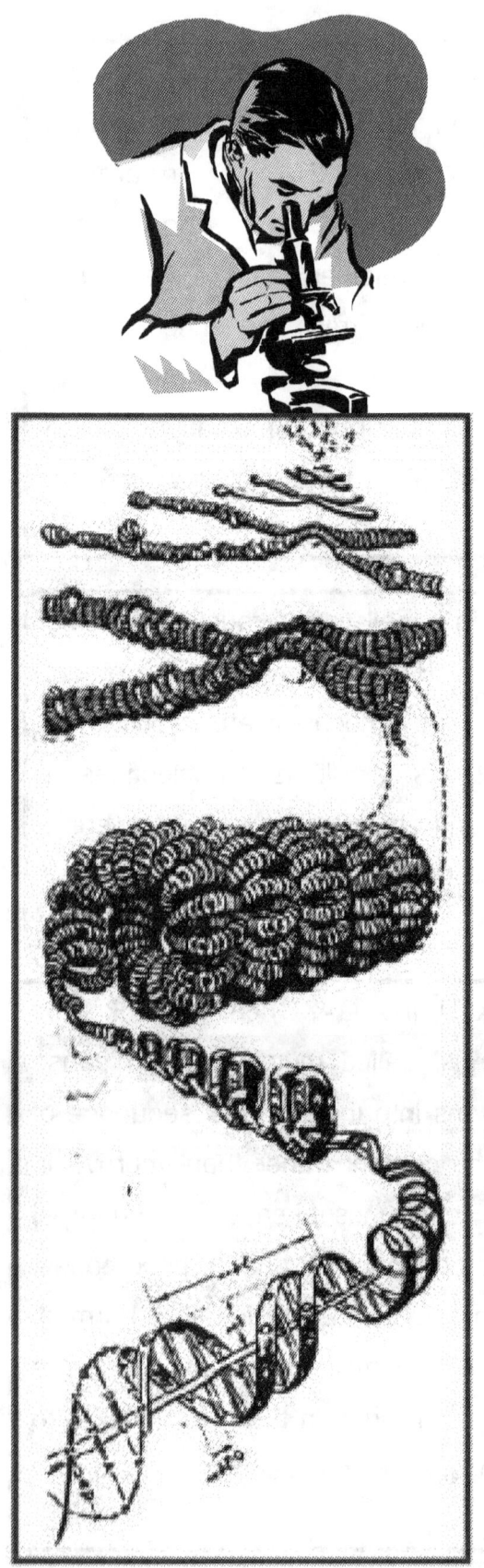

The question may come to your mind: How is it possible to get a string of six thousand million combinations into a single cell and have any room left over for the cell to function?

The truth is that the track, or string, of base combinations would be extremely long, extending some two meters (or more than six feet) if we were to lay it out like a railroad track.

The secret is the double twisting helix spiral design of DNA. Another factor is the smallness of the six basic molecules that compose the network. By this design, the whole structure is tidily packed into a space not much more than several hundredths of a millimeter, about the dimension at which light wavelengths are bent.

Splitting the track into pieces known as chromosomes increases the packing efficiency further. In the human cell, there are 46 such pieces.

Is all of this design by chance? No, certainly it is not by chance, but by

Intelligent Design.

Scientists put it a little more directly: "Calculations have invariably shown that the probability of obtaining functionally sequenced bio-macromolecules at random is vanishingly small, even on a scale of billions of years. (Ref.39)

We have only considered the random occurrence of one of the necessary aspects of the primordial cell. In addition, the cell must not only win one lottery, but the first four lotteries *all at once* in order to function and reproduce.

Do you recall the mousetrap that must be complete before it could function? Remember that all fundamental aspects of the first cell must be present for the cell to live, grow, and reproduce.

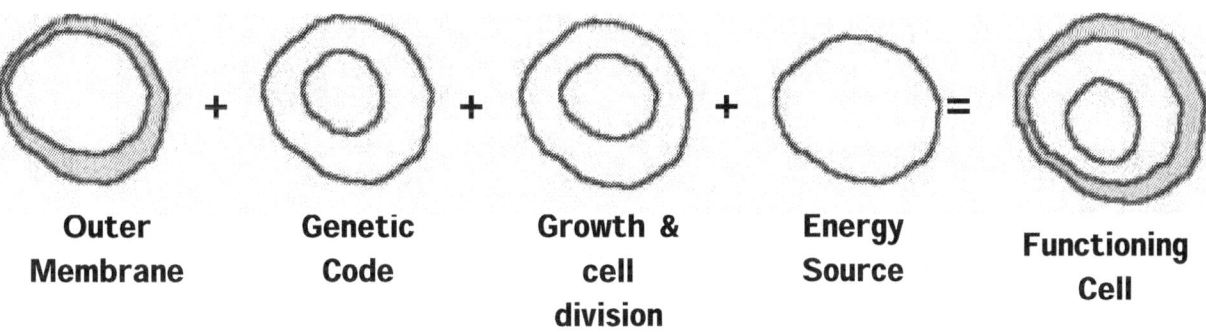

| Outer Membrane | Genetic Code | Growth & cell division | Energy Source | Functioning Cell |

We have considered just the chance formation of a single protein. A fundamental structure, such as the cell membrane is made up of many proteins and *all* of these must come together instantly to form a membranous envelope to hold the cell contents plus regulate the flow of nutrients and wastes in and out of the cell. Consider the three other fundamental structures that must be present for the cell to survive. The proposal of Darwinian Chance Determinism does not, could not account for such complexity. The clear alternative is Intelligent Design.

The theory of 'self-organization of first cell" has been repudiated by one of its early advocates, Dr. Dean Kenyon. Dr. Kenyon considers biological chance determinism as 'being both incompatible with empirical findings and theoretically incoherent." (Ref.24)

"The idea of Darwinian molecular evolution is not based upon science. There is no publication in the scientific literature . . . in journals, or books . . . that describes how molecular evolution of any real, complex, biochemical system either did occur or even might have occurred. There are assertions that such an evolution occurred, but absolutely none are supported by pertinent experiments or calculations. Since there is no authority on which to base claims of knowledge, it can reliably be said that the assertion of Darwinian molecular evolution is merely bluster."

Michael Behe, Ph.D. Assoc. Professor, Dept. of Biological Sciences, Lehigh University From: Intelligent Design Theory as a Tool for Analyzing Biochemical Systems, 1998. Mere Creation, Intervarsity Press, Downers Grove, Ill.

Despite total lack of evidence, either experimental or mathematical, the fantasy of the "self-organizing first cell" is being taught in our high schools as not only possible, but as actual fact.

The repudiation of a Creator goes far beyond the technical discussions of biologists in their scientific journals. It effects our young people in the manner described by the Cambridge Professor Herbert Butterfield:

" . . . belief in God determines:

☆ one's interpretation of all human history,

☆ the framework in which one views his/her fellow man,

☆ the existence of Earth and the entire Universe."

Imagine two scientists standing in awe before the great solid sculptured heads of the American presidents at Mt. Rushmore National Monument. One of them says, "What a remarkable likeness these heads have to the great men of America!" The other replies, "Yes, just think of how through the years, the forces of wind, rain and lightning have carved these perfect likenesses out of solid rock!"

Foolish? No, the actual chances of these heads to have occurred by pure random forces of nature ARE FAR BETTER than for life to have started spontaneously by chance over a period of billions of years, as proposed by macro-evolution.

Remember the statement by Harvard Professor of Biology, George Wald:

". . . However improbable we regard this event [the start of all life],
or any of the steps which it involves, given enough time it will
almost certainly happen. . ." (Ref. 48.1)

This statement appears in high school Biology books . . . but what <u>does not appear</u> is the the *RETRACTION* and *APOLOGY* by <u>*Scientific American*</u> for its publication in 1979, the only ever made by any scientific journal of a Nobel Laureate's writing. The following is the retraction published by Scientific American in 1979:

". . . .George Wald was completely wrong in his main thesis of forming a
biological cell by chance combinations of organic compounds. . . . Harold
Morowitz has shown that this process would require more time than the
Universe has been in existence." (Ref.14.1)

How misleading to continue to publish a totally FALSE concept when it has been retracted by the scientific journal that published it!

Psalm 8

O, Lord, our Lord,
how majestic is your name in all the earth!
. . .
When I consider your heavens,
the work of your fingers,
the moon and the stars,
which you have set in place,
what is man that you are mindful of him,
the son of man that you care for him?
You made him a little lower than the
heavenly beings
and crowned him with glory and honor.

You made him ruler over the works of
your hands;
you put everything under his feet
all flocks and herds,
and the beasts of the field,
the birds of the air,
and the fish of the sea,
all that swim the paths of the seas.

O Lord, our Lord,
How majestic is your name in all the earth!

(NIV)

References

1. Alberts, B., D. Bray, J. Lewis, M. Ralf, K. Roberts, and J.D. Watson. 1983. *Molecular biology of the cell.* New York: Garland,pp.94-141.

2. Allegre, C. & Schneider, S. 1994, "The Evolution of the Earth", *Scientific American,* Oct.

3. Behe, M., Ph.D Assoc. Prof: Dept of Biological Sciences, Lehigh Univ.

    ----- 1993. "Molecular machines: Experimental support for the design inference". Unpublished paper presented to ASA Intelligent Design Symposium, August 9.

    -----.1994. "Experimental support for regarding functional classes of proteins to be highly isolated from each other". In *Darwinism: Science or philosophy,* ed. J. Buell and G. Hearn, 60-71. Dallas: Foundation for Thought and Ethics

    -----.1996. *Darwin's black box: The biochemical challenge to evolution.* New York: Free Press.

    ----- 1998."Intelligent Design Theory as a Tool for Analyzing Biochemical Systems". In *Mere Creation*, InterVarsity Press, Downers Grove, Ill.

4. Bolte, M. & Hogan, J. 1995 "Conflict over the age of the Universe", in *Nature*, 376-402,

5. Borel, E. 1962. *Probabilities and life.* Translated by M. Baudin. New York: Dover Publications.

6. Bowler, P.1988. *The non-Darwinian revolution: Reinterpreting historical myth.* Baltimore: Johns Hopkins University Press.

7. Bowring, et al. 1993. "Calibrating Rates of Early Cambrian Evolution", in *Science* 261:1293.

8. Britter, R. & Kohne, 1970 "Repeated Segments of DNA" *Scientific American,* April.

9. Brock, D. L. 1992. *Our universe: Accident or design ?* Wits, South Africa: Star Watch

9.1 Bronowski, J. 1973 The Ascent of Man. Boston/Toronto: Little, Brown & Co. 59-70.

10. Canfield,D. & Teshe, A. 1996. Late Proterozoic Rise in Atmospheric Oxygen Concentration, In *Nature, 382:127*

11. Crick, F.   1968. "The origin of the genetic code". *Journal of Molecular Biology* 38:367-79.

-----1981. *Life itself.* New York: Simon and Schuster.

12. Dembski, W. A. 1998. *The design inference: Eliminating chance through small probabilities.* Cambridge: Cambridge University Press.

13. Denton, M. 1986. *Evolution: A theory in Crisis.* London: Adler and Adler.

14. Eddington, A. 1987. *Space, time and gravitation.* 1920. Reprint, Cambridge: Cambridge University Press.

14.1 Folsome, C. 1979, "Life: Origin and Evolution", *Scientific American Spl. Pub. 1979*

15. Forbes, G. 1993. "Time, events and modality". In *The philosophy of time,* ed. R. Le Poidevin and M MacBeath, 80-95. Oxford: Oxford University Press.

15.1 Ghyka, M., *The Geometry of Art and Life,* Dover Publications, Inc. (977, p.89)

15.2 Gitt, W. 1989. Information The third fundamental quantity. *Siemens Review* 56 (6):2-7.

16. Gould, S. Nov. 1992, "The Evolutionary Life on Earth". In *Scientific American.*

16.1 Green, B. 1999, *The Elegant* Universe, Vantage Books, NY: Random House (pp. 15-17, 144-151)

17. Harada, K., and S. Fox. 1964. "Thermal synthesis of amino acids from a postulated primitive terrestrial atmosphere". In *Nature* 201:335-37.

18. Hawking, S. 1988. *A Brief History of Time.* New York: Bantam.

19. Hawking, S., and R. Penrose. 1996. *The Nature of Space and Time.* Princeton, NJ.: Princeton University Press.

20. Healey, J. 1990. *The Early Alphabet,* British Museum Publications, London.

21. Hoyle, Fred 1981, *The Nature of the Universe.* N.Y. Harper

-----1982, *Astronomy and Cosmology,* W.H.Freeman, pp.540-541,2

22. Johnson, P. E. 1991. *Darwin on trial.* Washington, D.C.: Regnery Gateway

-----1995. *Reason in the balance: The case against naturalism in science, law and education.* Downers Grove, Ill.: InterVarsity Press.

23. Kaiser, J. 1994, "A New Theory of the Insect Wing Origins", In *Science* 266:363.

24. Kenyon, Dean 1987, "A Statistical Examination of Self-Ordering of Amino Acids in Proteins". In *Origins of Life, An Evolution of the Biosphere,* 18:135-42, Kok, Taylor & Bradley

25. Kerr, R. 1993. "Evolution's Big Bang gets even more explosive". In *Science* 261:1274

26. Kimber, Gordon, and Athwal, R.S 1972. "A Reassessment of the Course of Evolution of Wheat", *Proceedings of the National Academy of Science,* 69 no.4, pp912-15, April

27. Kornberg, A., and T. A. Baker. 1992. *DNA replication.* 2nd ed. New York: Freeman

28. Maimonides, 1190. *Guide for the Perplexed,* 1:7

28.1 Meyer, S.C. 1993. A Scopes Trial for the '90s. *The Wall Street Journal, December 6, A14*

29. Nahmanides. 1250, commentary on *Genesis* 1:1-2

30. Nahmanides. 1250, commentary on *Genesis* 1:12

31. Nahmanides. 1250, commentary on *Genesis* 1:21, 1:27

32. Nahmanides. 1250, commentary on *Genesis* 2:7

33. Nahmanides. 1250, commentary on *Genesis* 6:4

34. Nahmanides. 1250, commentary on *Exodus* 7-10

35. Nash, M. 1995. "When Life Exploded", *Time Magazine,* Dec.4, 1995

36. Oparin, A.I. 1938. *The origin of life.* Translated by S. Morgulis. New York: Macmillan.
    ----- 1968. *Genesis and evolutionary development of life.* Translated by E. Maass. New York: Academic Press.

37. Peebles, P. J. E. 1993. *Principles of physical cosmology.* Princeton, NJ.: Princeton; UniversityPress.

38. Penrose, R. 1989. *The Emperor's New Mind,* Penguin Books, N.Y.

39. Prigogine, I. & Nicolis, G. 1977. *Self organization in nonequilibrium systems.* New York: Wiley.

40. Rashi, commentary on Talmud Sanhedrin 97A 1030 C.E.; Leviticus Rabba 29:1

41. Ross, Hugh. 1995a. *The Creator and the Cosmos.* 2nd ed. Colorado Springs, Colo: NavPress.

    ------ 1998. "Big Bang Model Refined by Fire". In *Mere Creation,* Ed: William Dembski, 363-384. Downers Grove, Ill. InterVarsity Press.

42. Schroeder, Gerald 1997. *The Science of God.* The Free Press, New York pp98-99.

43. Scott, A. 1986. *The Creation of Life.* Oxford: Oxford University Press.

44. Silk, J. 1989. *The Big Bang,* W.H.Freeman, New York, p.72

45. Talmud Keliim 8:5

46. Talmud Keliim 12A; Rashi

47. Thaxton, C. B., and W. L. Bradley. 1994. "Information and the origin of life". In *The creation hypothesis,* ed. J. P. Moreland, 173-210. Downers Grove, Ill.: Inter\\\'arsity Press.

48. Tipler, F. J., and J. D. Barrow. 1985 *The Anthropic cosmological principle.* York: Oxford University Press.

49. Walker, C.B. 1989. *Reading the Past:Cuneiform,* British Museum Publications, London.

50. Watson, J., and E Crick. 1953. "A structure for deoxyribose nucleic acid". In *Nature* 171:737-38.

51. Weinberg, S. 1977. *The First Three Minutes.* Basic Books, New York

    -----1992. *Dreams of a final theory.* New York; Pantheon.

52. Wells, J. 1998. "Unseating naturalism: Recent insights from developmental biology". In *Mere Creation,* ed. W. A. Dembski, chap. 2. Downers Grove, Ill.: InterVarsity Press.

53. Wolfe, S. L. 1993. *Molecular and cellular biology.* Belmont, Calif.: Wadsworth.

54. Yockey, H. P 1977. "A calculation of the probability of spontaneous biogenesis by information theory". In *Journal of Theoretical Biology* 67:377-98.

55. Young, J.Z. 1971 *An Introduction to Study of Man,* Oxford University Press.

Who is Paul Garnett?

Paul Garnett received his training in science and education at several universities here in the United States, and he completed his graduate study in the field of Human Behavior at Oxford University, England.

Dr. Garnett founded and was Director of the first Montessori school in the State of Oregon and then went on to establish three other Montessori schools before he moved to the Hawaii State Department of Education. In Hawaii, he developed a non-verbal IQ test for individuals who do not speak Standard English. He also wrote a text now being widely used throughout America and Canada, entitled: *Investigating Morals and Ethics in Today's Society* (Simon & Schuster 1987).

Dr. Garnett is available for seminars, lectures & presentations to teachers, students, church groups and those interested in Creation. Contact:

Dr. Paul D. Garnett
Unity Research Institute
4840 26th Court S.
St. Petersburg, FL 33712
FAX: (727) 906-8148